D0622752

EUCLID
and His
MODERN RIVALS

EUCLID, BOOK I.

Arranged in Logical Sequence.

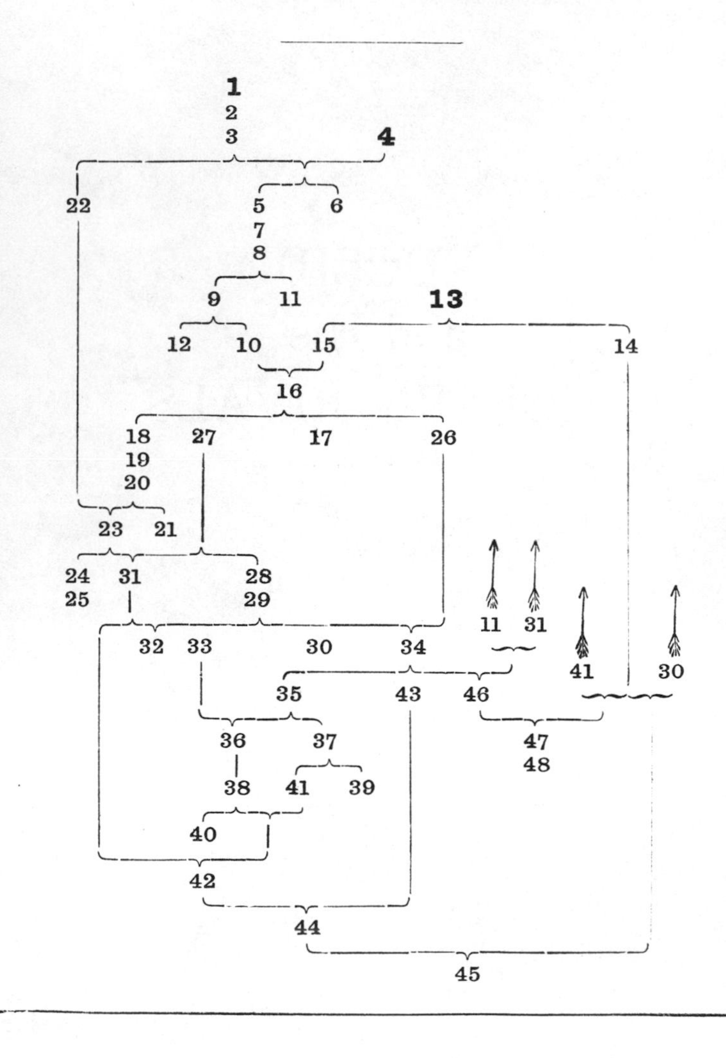

EUCLID and His MODERN RIVALS

by
Lewis Carroll
(Charles L. Dodgson, M.A.)

With a new Introduction by
H. S. M. COXETER
Professor of Mathematics,
University of Toronto

Dover Publications, Inc.
New York

Published in Canada by General Publishing Company, Ltd., 30 Lesmill Road, Don Mills, Toronto, Ontario.
Published in the United Kingdom by Constable and Company, Ltd., 10 Orange Street, London WC 2.

This Dover edition, first published in 1973, is an unabridged republication of the second edition of the work, published in London by Macmillan and Co. in 1885. A new introduction by H.S.M. Coxeter has been added in the present edition.

International Standard Book Number: 0-486-22968-8
Library of Congress Catalog Card Number: 73-80346

Manufactured in the United States of America
Dover Publications, Inc.
180 Varick Street
New York, N. Y. 10014

Dedicated

to

the memory

of

Euclid

INTRODUCTION
TO THE DOVER EDITION

ECHOING Gerolamo Saccheri, whose *Euclides ab omni naevo vindicatus* was published in 1733, Dodgson writes (in the Preface to First Edition) "of the great cause which I have at heart—the vindication of Euclid's masterpiece." To make this project entertaining, he expresses it as a drama in which Euclid converses with Minos and Rhadamanthus, two of the three judges in Hades. Herr Niemand, "the Phantasm of a German Professor," appears on page 17, as spokesman for the 13 "modern rivals" who were trying to make the treatment of geometry more rigorous, or more palatable, by revising the definitions and rearranging the propositions. One by one, these rivals are ridiculed and put to shame. Time has justified Dodgson's scorn, for all save Legendre and Peirce have been forgotten. A. M. Legendre's *Eléments de Géométrie,* "though well suited for advanced students, is not so for beginners" (p. 59). According to Professor Benjamin Peirce (spelled "Pierce" in the text) of Harvard (pp. 144-147), "Parallel Lines are straight Lines which have the same Direction." And how did he define Direction? "The Direction of a Line in any part is the direction of a point at that part from the next preceding point of the Line."

Were Dodgson still alive, he would be equally indignant about a new generation of textbook writers, who try to make geometry easier by introducing redundant "postulates" or "assumptions." On page 57 we read, ". . . it is a generally admitted principle that, at least in dealing with beginners, we ought not to take as axiomatic any Theorem which can be proved by the Axioms we already possess." There is much to be said for his standpoint that the degree of rigor in Euclid's *Elements* is just right for high school: a modern axiomatic treatment (such as H. G. Forder, *The Foundations of Euclidean Geometry,* Dover, 1958) should be left for mature students in universities.

In *The summing-up* (p. 225) the departing ghost of Euclid pleads: "Let me carry with me the hope that I have convinced you of the importance, if not the necessity, of retaining my order and numbering, and my method of treating straight Lines, angles, right angles, and (most especially) Parallels." And again (p. 11): "The Propositions have been known by those numbers for two thousand years; . . . and some of them, I.5 and I.47 for instance— 'the Asses' Bridge' and 'the Windmill'— are now historical characters, and their nicknames are 'familiar as household words.' "

The book is by no means all Dodgson the serious teacher: every now and then Lewis Carroll appears with some unexpected analogy or outrageous pun. (See, for instance, pages 48 and 119.) And there are many quotable remarks, such as ". . . analogies give to Geometry much of its beauty" (p. 221). (For some further observations on these aspects of his work, see P. L. Heath's delightful article "Carroll, Lewis," in the *Encyclopaedia of Philosophy.*)

On pages 28–36 we find "TABLE I, *Containing twenty Propositions, of which some are undisputed Axioms, and the rest real and valid Theorems, deducible from undisputed Axioms*" and "TABLE II, *Containing eighteen Propositions of which no one is an undisputed Axiom, but all are real and valid Theorems, which, though not deducible from undisputed Axioms, are such that, if any one be admitted as an Axiom, the rest can be proved.*" A few years later, Dodgson would doubtless have included, in the latter table, a nineteenth item (see "A New Theory of Parallels," which is Part I of *Curiosa Mathematica,* 3rd ed., London, 1890, p. 14): *In every circle, the inscribed equilateral tetragon is greater than any one of the segments which lie outside it.*

Tables I and II are his nearest approach to the subject of non-Euclidean geometry. In these days of enlightenment we find it difficult to realize that, 100 years ago, Professors Arthur Cayley of Cambridge and W. K. Clifford of London may well have been the only Englishmen who understood the philosophical revolution that had been instigated by Gauss, Bolyai and Lobachevsky, some 50 or 60 years earlier. One is tempted to speculate on what might have happened if Cayley or Clifford had met Dodgson and convinced him that there is a logically consistent "hyperbolic" geometry in which the "absolute" propositions in Table I still hold while all the statements in Table II are false (and the nineteenth proposition fails for any sufficiently large circle). In his *Sylvie and Bruno Concluded* (London, 1893, pp. 100–104) the real projective plane is represented as a "Purse of Fortunatus" made by sewing together three square handkerchiefs.

The same easy style and fertile imagination, applied to the infinite hyperbolic plane, would surely have produced a thrilling exploration of this new Wonderland.

H. S. M. COXETER

Toronto, Canada
March, 1973

PREFACE TO SECOND EDITION.

The only new features, worth mentioning, in the second edition, are the substitution of words for the symbols introduced in the first edition, and one additional review—of Mr. Henrici, to whom, if it should appear to him that I have at all exceeded the limits of fair criticism, I beg to tender my sincerest apologies.

C. L. D.

Ch. Ch. 1885.

PREFACE TO FIRST EDITION.

'ridentem dicere verum
Quid vetat?'

The object of this little book is to furnish evidence, first, that it is essential, for the purpose of teaching or examining in elementary Geometry, to employ one text-book only; secondly, that there are strong *a priori* reasons for retaining, in all its main features, and specially in its sequence and numbering of Propositions and in its treatment of Parallels, the Manual of Euclid; and thirdly, that no sufficient reasons have yet been shown for abandoning it in favour of any one of the modern Manuals which have been offered as substitutes.

It is presented in a dramatic form, partly because it seemed a better way of exhibiting in alternation the arguments on the two sides of the question; partly that I might feel myself at liberty to treat it in a rather lighter style than would have suited an essay, and thus to make it a little less tedious and a little more acceptable to unscientific readers.

In one respect this book is an experiment, and may chance to prove a failure: I mean that I have not thought it necessary to maintain throughout the gravity of style which scientific writers usually affect, and which has somehow come to be regarded as an 'inseparable accident' of scientific teaching. I never could quite see the reasonableness of this immemorial law: subjects there are, no doubt, which are in their essence too serious to admit of any lightness of treatment—but I cannot recognise Geometry as one of them. Nevertheless it will, I trust, be found that I have permitted myself a glimpse of the comic side of things only at fitting seasons, when the tired reader might well crave a moment's breathing-space, and not on any occasion where it could endanger the continuity of a line of argument.

Pitying friends have warned me of the fate upon which I am rushing: they have predicted that, in thus abandoning the dignity of a scientific writer, I shall alienate the sympathies of all true scientific readers, who will regard the book as a mere *jeu d'esprit*, and will not trouble themselves to look for any serious argument in it. But it must be borne in mind that, if there is a Scylla before me, there is also a Charybdis—and that, in my fear of being read as a jest, I may incur the darker destiny of not being read at all.

In furtherance of the great cause which I have at heart —the vindication of Euclid's masterpiece—I am content to run some risk; thinking it far better that the purchaser of this little book should *read* it, though it be with a smile, than that, with the deepest conviction of its seriousness of purpose, he should leave it unopened on the shelf.

To all the authors, who are here reviewed, I beg to tender my sincerest apologies, if I shall be found to have transgressed, in any instance, the limits of fair criticism. To Mr. Wilson especially such apology is due—partly because I have criticised his book at great length and with no sparing hand—partly because it may well be deemed an impertinence in one, whose line of study has been chiefly in the lower branches of Mathematics, to dare to pronounce any opinion at all on the work of a Senior Wrangler. Nor should I thus dare, if it entailed my following him up 'yonder mountain height' which *he* has scaled, but which *I* can only gaze at from a distance: it is only when he ceases 'to move so near the heavens,' and comes down into the lower regions of Elementary Geometry, which I have been teaching for nearly five-and-twenty years, that I feel sufficiently familiar with the matter in hand to venture to speak.

Let me take this opportunity of expressing my gratitude, first to Mr. Todhunter, for allowing me to quote *ad libitum* from the very interesting Essay on Elementary Geometry, which is included in his volume entitled 'The Conflict of Studies, and other Essays on subjects connected with Education,' and also to reproduce some of the beautiful diagrams from his edition of Euclid; secondly, to the Editor of the Athenæum, for giving me a similar permission with regard to a review of Mr. Wilson's Geometry, written by the late Professor De Morgan, which appeared in that journal, July 18, 1868.

C. L. D.

Ch. Ch. 1879.

ARGUMENT OF DRAMA.

ACT I.

Preliminaries to examination of Modern Rivals.

Scene I.

[Minos *and* Rhadamanthus.]

SCENE II.

[MINOS *and* EUCLID.]

§ 1. A priori *reasons for retaining Euclid's Manual.*

§ 2. *Method of procedure in examining Modern Rivals.*

§ 3. *The combination, or separation, of Problems and Theorems.*

§ 4. *Syllabus of propositions relating to Pairs of Lines.*

§ 5. *Playfair's Axiom.*

§ 6. *Principle of Superposition.*

SCENE III.

Treatment of Parallels by angles made with transversals.

COOLEY.

SCENE IV.

Treatment of Parallels by equidistances.

CUTHBERTSON.

Scene V.

Treatment of Parallels by revolving lines.

Henrici.

Scene VI.

Treatment of Parallels by direction.

§ 1. Wilson.

§ 2. PIERCE.

§ 3. WILLOCK.

ACT III.

Manuals which adopt Euclid's treatment of Parallels.

SCENE I.

§ 1.

§ 2. CHAUVENET.

§ 3. LOOMIS.

§ 4. MORELL.

§ 5. REYNOLDS.

§ 6. WRIGHT.

SCENE II.

§ 1. SYLLABUS OF THE ASSOCIATION FOR THE IMPROVEMENT OF GEOMETRICAL TEACHING.

§ 2. WILSON'S 'SYLLABUS'-MANUAL.

Of 73 Propositions of Euclid, this Manual has
14 omitted;
43 done as in Euclid;
10 done by new but objectionable methods, viz.—
1 illogical;
1 'hypothetical construction';
2 needlessly using 'superposition';
2 algebraical;
4 omitting the diagonals of Euc. II.;
6 done by new and admissible methods.

ACT IV.

[MINOS *and* EUCLID.]

Manual of Euclid.

§ 1. *Treatment of Pairs of Lines.*

§ 2. *Euclid's constructions.*

§ 3. *Euclid's demonstrations.*

§ 4. *Euclid's style.*

§ 5. *Euclid's treatment of Lines and Angles.*

§ 6. *Omissions, alterations, and additions, suggested by Modern Rivals.*

§ 7. *The summing-up.*

APPENDICES.

IV.

ACT I.

Scene I.

'Confusion worse confounded.'

[*Scene, a College study. Time, midnight.* Minos *discovered seated between two gigantic piles of manuscripts. Ever and anon he takes a paper from one heap, reads it, makes an entry in a book, and with a weary sigh transfers it to the other heap. His hair, from much running of fingers through it, radiates in all directions, and surrounds his head like a halo of glory, or like the second Corollary of* Euc. I. 32. *Over one paper he ponders gloomily, and at length breaks out in a passionate soliloquy.*]

Min. So, my friend! *That's* the way you prove I. 19, is it? Assuming I. 20? Cool, refreshingly cool! But stop a bit! Perhaps he doesn't 'declare to win' on Euclid. Let's see. Ah, just so! 'Legendre,' of course! Well, I suppose I must give him full marks for it: what's the question worth?—Wait a bit, though! Where's his paper of yesterday? I've a very decided impression he was all for 'Euclid' then: and I know the paper had I. 20 in it.

. . . Ah, here it is! 'I think we do know the sweet Roman hand.' Here's the Proposition, as large as life, and proved by I. 19. 'Now, infidel, I have thee on the hip!' You shall have such a sweet thing to do in *vivâ-voce*, my very dear friend! You shall have the two Propositions together, and take them in any order you like. It's my profound conviction that you don't know how to prove either of them without the other. They'll have to introduce each other, like Messrs. Pyke and Pluck. But what fearful confusion the whole subject is getting into! (*Knocking heard.*) Come in!

Enter RHADAMANTHUS.

Rhad. I say! Are we bound to mark an answer that's a clear logical fallacy?

Min. Of course you are—with that peculiar mark which cricketers call 'a duck's egg,' and thermometers 'zero.'

Rhad. Well, just listen to this proof of I. 29.

Reads.

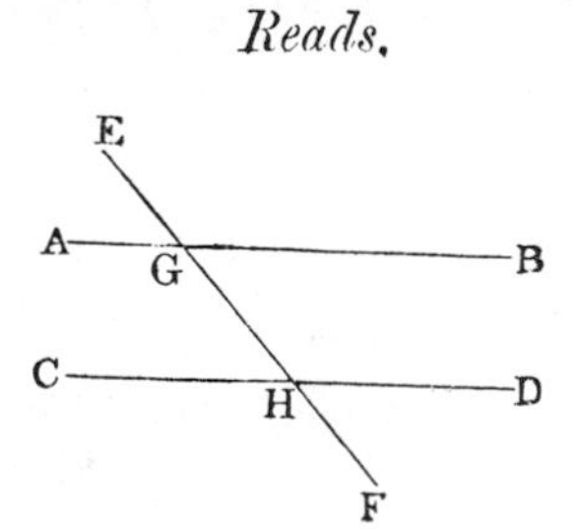

'Let *EF* meet the two parallel Lines *AB*, *CD*, in the points *GH*. The alternate angles *AGH*, *GHD*, shall be equal.

'For AGH and EGB are equal because vertically opposite, and EGB is also equal to GHD (Definition 9); therefore AGH is equal to GHD; but these are alternate angles.'

Did you ever hear anything like that for calm assumption?

Min. What does the miscreant mean by 'Definition 9'?

Rhad. Oh, that's the grandest of all! You must listen to that bit too. There's a reference at the foot of the page to 'Cooley.' So I hunted up Mr. Cooley among the heaps of Geometries they've sent me—(by the way, I wonder if they've sent *you* the full lot? Forty-five were left in my rooms to-day, and ten of them I'd never even heard of till to-day!)—well, as I was saying, I looked up Cooley, and here's the Definition.

Reads.

'Right Lines are said to be parallel when they are equally and similarly inclined to the same right Line, or make equal angles with it towards the same side.'

Min. That is very soothing. So far as I can make it out, Mr. Cooley quietly assumes that a Pair of Lines, which make equal angles with *one* Line, do so with *all* Lines. He might just as well say that a young lady, who was inclined to *one* young man, was 'equally and similarly inclined' to *all* young men!

Rhad. She might 'make equal angling' with them all, anyhow. But, seriously, what are we to do with Cooley?

Min. (*thoughtfully*) Well, if we had him in the Schools, I *think* we should pluck him.

Rhad. But as to this answer?

Min. Oh, give it full marks! What have we to do with logic, or truth, or falsehood, or right, or wrong? 'We are but markers of a larger growth'—only that *we* have to mark foul strokes, which a respectable billiard-marker doesn't do, as a general rule!

Rhad. There's one thing more I want you to look at. Here's a man who puts 'Wilson' at the top of his paper, and proves Euc. I. 32 from first principles, it seems to me, without using any other Theorem at all.

Min. The thing sounds impossible.

Rhad. So *I* should have said. Here's the proof.

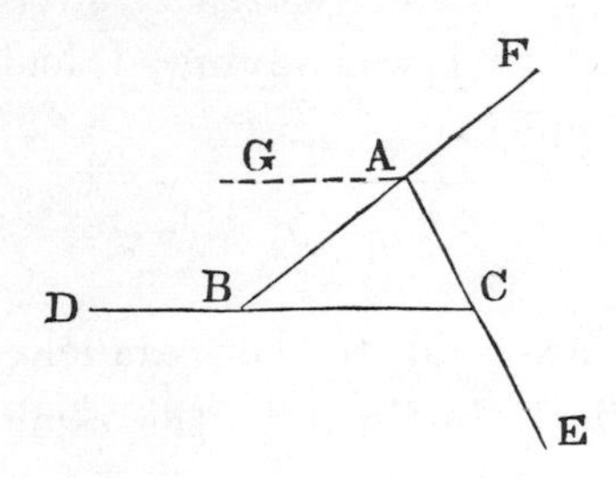

'Slide ∠ *DBA* along *BF* into position *GAF*, *GA* having same direction as *DC* (Ax. 9); similarly slide ∠ *BCE* along *AE* into position *GAC*. Then the ext. ∠s = *CAF*, *FAG*, *GAC* = one revolution = two straight ∠s. But the ext. and int. ∠s = 3 straight ∠s. Therefore the int. ∠s = one straight ∠ = 2 right angles. Q. E. D.'

I'm not well up in 'Wilson': but surely he doesn't beg the whole question of Parallels in one axiom like this!

Min. Well, no. There's a Theorem and a Corollary. But this is a sharp man: he has seen that the Axiom does just as well by itself. Did you ever see one of those conjurers bring a globe of live fish out of a pocket-handkerchief? That's the kind of thing we have in Modern Geometry. A man stands before you with nothing but an Axiom in his hands. He rolls up his sleeves. 'Observe, gentlemen, I have nothing concealed. There is no deception!' And the next moment you have a complete Theorem, Q. E. D. and all!

Rhad. Well, so far as *I* can see, the proof's worth nothing. What am I to mark it?

Min. Full marks: we *must* accept it. Why, my good fellow, I'm getting into that state of mind, I'm ready to mark *any* thing and *any* body. If the Ghost in Hamlet came up this minute and said 'Mark me!' I should say 'I will! Hand in your papers!'

Rhad. Ah, it's all very well to chaff, but it's enough to drive a man wild, to have to mark all this rubbish! Well, good night! I must get back to my work. [*Exit.*

Min. (*indistinctly*) I'll just take forty winks, and—

(*Snores.*)

ACT I.

SCENE II.

Οὐκ ἀγαθὸν πολυκοιρανίη· εἷς κοίρανος ἔστω,
Εἷς βασιλεύς.

[MINOS *sleeping: to him enter the Phantasm of* EUCLID. MINOS *opens his eyes and regards him with a blank and stony gaze, without betraying the slightest surprise or even interest.*]

---

§ 1. A priori *reasons for retaining Euclid's Manual.*

Euc. Now what is it you really require in a Manual of Geometry?

Min. Excuse me, but—with all respect to a shade whose name I have reverenced from early boyhood—is not that *rather* an abrupt way of starting a conversation? Remember, we are twenty centuries apart in history, and consequently have never had a personal interview till now. Surely a few preliminary remarks—

Euc. Centuries are long, my good sir, but *my* time to-night is short: and I never was a man of many words. So kindly waive all ceremony and answer my question.

Min. Well, so far as I can answer a question that comes upon me so suddenly, I should say—a book that will exercise the learner in habits of clear definite conception, and enable him to test the logical value of a scientific argument.

Euc. You do *not* require, then, a complete repertory of Geometrical truth?

Min. Certainly not. It is the *ἐνέργεια* rather than the *ἔργον* that we need here.

Euc. And yet many of my Modern Rivals have thus attempted to improve upon me—by filling up what they took to be my omissions.

Min. I doubt if they are much nearer to completeness themselves.

Euc. I doubt it too. It is, I think, a friend of yours who has amused himself by tabulating the various Theorems which might be enunciated in the single subject of Pairs of Lines. How many did he make them out to be?

Min. About two hundred and fifty, I believe.

Euc. At that rate, there would probably be, within the limits of my First Book, about how many?

Min. A thousand, at least.

Euc. What a popular school-book it will be! How boys will bless the name of the writer who first brings out the complete thousand!

Min. I think your Manual is fully long enough already for all possible purposes of teaching. It is not in the region of new matter that you need fear your Modern

Rivals: it is in *quality*, not in *quantity*, that they claim to supersede you. Your methods of proof, so they assert, are antiquated, and worthless as compared with the new lights.

Euc. It is to that very point that I now propose to address myself: and, as we are to discuss this matter mainly with reference to the wants of *beginners*, we may as well limit our discussion to the subject-matter of Books I and II.

Min. I am quite of that opinion.

Euc. The first point to settle is whether, for purposes of teaching and examining, you desire to have one fixed logical sequence of Propositions, or would allow the use of conflicting sequences, so that one candidate in an examination might use *X* to prove *Y*, and another use *Y* to prove *X*—or even that the same candidate might offer *both* proofs, thus 'arguing in a circle.'

Min. A very eminent Modern Rival of yours, Mr. Wilson, seems to think that no such fixed sequence is really necessary. He says (in his Preface, p. 10) 'Geometry when treated as a science, treated inartificially, falls into a certain order from which there can be no very wide departure; and the manuals of Geometry will not differ from one another nearly so widely as the manuals of algebra or chemistry; yet it is not difficult to examine in algebra and chemistry.'

Euc. Books may differ very 'widely' without differing in logical sequence—the only kind of difference which could bring two text-books into such hopeless collision that the one or the other would have to be abandoned.

Let me give you a few instances of conflicting logical sequences in Geometry. Legendre proves my Prop. 5 by Prop. 8, 18 by 19, 19 by 20, 27 by 28, 29 by 32. Cuthbertson proves 37 by 41. Reynolds proves 5 by 20. When Mr. Wilson has produced similarly conflicting sequences in the manuals of algebra or chemistry, we may then compare the subjects: till then, his remark is quite irrelevant to the question.

Min. I do not think he will be able to do so: indeed there are very few logical chains *at all* in those subjects—most of the Propositions being proved from first principles. I think I may grant at once that it is essential to have *one* definite logical sequence, however many manuals we employ: to use the words of another of your Rivals, Mr. Cuthbertson (Pref. p. viii.), 'enormous inconvenience would arise in conducting examinations with no recognised sequence of Propositions.' This however applies to *logical* sequences only, such as your Props. 13, 15, 16, 18, 19, 20, 21, which form a continuous chain. There are many Propositions whose place in a manual would be partly arbitrary. Your Prop. 8, for instance, is not wanted till we come to Prop. 48, so that it might occupy any intermediate position, without involving risk of circular argument.

Euc. Now, in order to secure this uniform logical sequence, we should require to know, as to any particular Proposition, what other Propositions were its logical descendants, so that we might avoid using any of these in proving it?

Min. Exactly so.

Euc. We might of course give this information by attaching to each enunciation references to its logical descendants: but this would be a very cumbrous plan. A better way would be to give them in the form of a genealogy, but this would be very bulky if the enunciations themselves were inserted: so that it would be desirable to have numbers to distinguish the enunciations. In that case (supposing *my* logical sequence to be adopted) the genealogy would stand thus:—(*see Frontispiece*).

Min. Would it not be enough to publish an arranged list (which would be all the better if numbered also), and to enact that no Proposition should be used to prove any of its predecessors?

Euc. That would hamper the writers of manuals very much more than the genealogy would. Suppose, for instance, that you adopted, in the list, the order of Theorems in my First Book, and that a writer wished to prove Prop. 8 by Prop. 47: this would not interfere with my logical sequence, and yet your list would bar him from doing so.

Min. But we might place 8 close before 48, and he would then be free to do as you suggest.

Euc. And suppose some other writer wished to prove 24 by 8?

Min. I see now that any single list must necessarily prevent many possible arrangements which would not conflict with the agreed-on logical sequence. And yet this is what the Committee of the Association for the Improvement of Geometrical Teaching have approved of, namely, 'a standard sequence for examination purposes,' and what the Association have published in their 'Syllabus of Plane Geometry.'

Euc. I think they have overlooked the fact that they are enacting many more sequences, as binding on writers, than the one logical sequence which they desire to secure. Their 'standard sequence' would be fitly replaced by a 'standard genealogy.' But in any case we are agreed that it is desirable to have, besides a standard logical sequence, a standard list of enunciations, numbered for reference?

Min. We are.

Euc. The next point to settle is, *what* sequence and numbering to adopt. You will allow, I think, that there are strong *a priori* reasons for retaining *my* numbers. The Propositions have been known by those numbers for two thousand years; they have been referred to, probably, by hundreds of writers—in many cases by the numbers only, without the enunciations: and some of them, I. 5 and I. 47 for instance—'the Asses' Bridge' and 'the Windmill'—are now historical characters, and their nicknames are 'familiar as household words.'

Min. Even if no better *sequence* than yours could be found, it might still be urged that a new set of *numbers* must be adopted, in order to introduce, in their proper places, some important Theorems which have been added to the subject since your time.

Euc. That want, if it were proved to exist, might, I think, be easily provided for without discarding my system of numbers. If you wished, for instance, to insert two new Propositions between I. 13 and I. 14, it would be far less inconvenient to call them 13 B and 13 C than to abandon the old numbers.

Min. I give up the objection.

Euc. You will allow then, I think, that my sequence and system of numbers should not be abandoned without good cause?

Min. Oh, certainly. And the *onus probandi* lies clearly on your Modern Rivals, and not on you.

Euc. Unless, then, it should appear that one of my Modern Rivals, whose logical sequence is incompatible with mine, is so decidedly better in his treatment of really important topics, as to make it worth while to suffer all the inconvenience of a change of numbers, you would not recognise his demand to supersede my Manual?

Min. On that point let me again quote Mr. Wilson. In his Preface, p. 15, he says, 'In a few years I hope that our leading mathematicians will have published, perhaps in concert, one or more text-books of Geometry, not inferior, to say the least, to those of France, and that they will supersede Euclid by the sheer force of superior merit.'

Euc. And I should be quite content to be so superseded. 'A fair field and no favour' is all I ask.

---

§ 2. *Method of procedure in examining Modern Rivals.*

Min. You wish me then to compare your book with those of your Modern Rivals?

Euc. Yes. But, in doing this, I must beg you to bear

in mind that a Modern Rival will not have proved his case if he only succeeds in showing

(1) that certain Propositions might with advantage be omitted (for this a teacher would be free to do, so long as he left the logical sequence complete);

or (2) that certain proofs might with advantage be changed for others (for these might be interpolated as 'alternative proofs');

or (3) that certain new Propositions are desirable (for these also might be interpolated, without altering the numbering of the existing Propositions).

All these matters will need to be fully considered hereafter, if you should decide that my Manual ought to be retained: but they do not constitute the evidence on which that decision should be based.

Min. *That*, I think, you have satisfactorily proved. But what *would* you consider to be sufficient grounds for abandoning your Manual in favour of another?

Euc. Many grave charges have been brought against my Manual; but, of all these, there are only *two* which I regard as *crucial* in this matter. The first concerns my arrangement of Problems and Theorems: the second my treatment of Parallels.

If it be agreed that Problems and Theorems ought to be treated separately, my system of numbering must of course be abandoned, and no reason will remain why my Manual should then be retained as a *whole*; which is the only point I am concerned with. This question you can, of course, settle on its own merits, without examining any of the new Manuals.

If, again, it be agreed that, in treating Parallels, some other method, *essentially* different from mine, ought to be adopted, I feel that, after so vital a change as that, involving (as no doubt it would) the abandonment of my sequence and system of numbering, the remainder of my Manual would not be worth fighting for, though portions of it might be embodied in the new Manual. To settle this question, you must, of course, examine one by one the new methods that have been proposed.

Min. You would not even ask to have your Manual retained as an alternative for the new one?

Euc. No. For I think it essential for purposes of teaching, that in treating this vital topic one uniform method should be adopted; and that this method should be the best possible (for it is almost inconceivable that two methods of treating it should be *equally* good). An alternative proof of a minor Proposition may fairly be inserted now and then as about equal in merit to the standard proof, and may make a desirable variety: but on this one vital point it seems essential that nothing but the best proof existing should be offered to the limited capacity of a learner. *Vacuis committere venis nil nisi lene decet.*

Min. I agree with you that we ought to have one system only, and that the best, for treating the subject of Parallels. But would you have me limit my examination of your 'Modern Rivals' to this single topic?

Euc. No. There are several other matters of so great importance, and admitting of so much variety of treatment, that it would be well to examine any method of dealing with them which differs much from mine—not with a view of substituting the new Manual for mine, but in order to make such changes in my proofs as may be thought desirable. There are other matters again, where changes have been suggested, which you ought to consider, but on *general* grounds, not by examining particular writers.

Let me enumerate what I conceive should be the subjects of your enquiry, arranged in order of importance.

(1) The combination, or separation, of Problems and Theorems.

(2) The treatment of Pairs of Lines, especially Parallels, for which various new methods have been suggested. These may be classified as involving—

(α) Infinite series: suggested by LEGENDRE.

(β) Angles made with transversals: COOLEY.

(γ) Equidistance: CUTHBERTSON.

(δ) Revolving Lines: HENRICI.

(ϵ) Direction: WILSON, PIERCE, WILLOCK.

(ζ) The substitution of 'Playfair's Axiom' for my Axiom 12.

If your decision, on these two crucial questions, be given in my favour, we may take it as settled, I think, that my Manual ought to be retained as a *whole:* how far it should be modified to suit modern requirements will be matter for further consideration.

(3) The principle of superposition.

(4) The use of diagonals in Book II.

These two are *general* questions, and will not need the examination of particular authors.

Besides this, it will be well, in order that your enquiry into the claims of my Modern Rivals may be as complete as possible, to review them one by one, with reference to their treatment of matters not already discussed, especially :—

(5) Right Lines.
(6) Angles, including right angles.
(7) Propositions of mine omitted.
(8) Propositions of mine treated by a new method.
(9) New Propositions.
(10) And you may as well conclude, in each case, with a general survey of the book, as to style, &c.

The following may be taken as a fairly complete catalogue of the books to be examined :—

1. Legendre.
2. Cooley.
3. Cuthbertson.
4. Henrici.
5. Wilson.
6. Pierce.
7. Willock.
8. Chauvenet.
9. Loomis.
10. Morell.
11. Reynolds.
12. Wright.
13. Wilson's 'Syllabus'-Manual.

You should also examine the Syllabus, published by the Association for the Improvement of Geometrical Teaching, on which the last-named Manual is based. Not that it can be considered as a '*Rival*'—in fact, it is not a text-book at all, but a mere list of enunciations—but because, first, it comes with an array of imposing names to recommend it, and secondly, it discards my system of numbers, so that its adoption, as a standard for examinations, would seriously interfere with the retention of my Manual as the standard text-book.

Now, of these questions, I shall be most happy to discuss with you the *general* ones (I mean questions 1, 2 (ζ), 3, and 4) before we conclude this interview: but, when it comes to criticising particular authors, I must leave you to yourself, to deal with them as best you can.

Min. It will be weary work to do it all alone. And yet I suppose you cannot, even with *your* supernatural powers, fetch me the authors themselves?

Euc. I dare not. The living human race is so strangely prejudiced. There is nothing men object to so emphatically as being transferred by ghosts from place to place. I cannot say they are consistent in this matter: they are for ever 'raising' or 'laying' us poor ghosts—we cannot even haunt a garret without having the parish at our heels, bent on making us change our quarters: whereas if *I* were to venture to move one single small boy—say to lift him by the hair of his head over only two or three houses, and to set him down safe and sound in a neighbour's garden—why, I give you my word, it would be the talk of the town for the next month!

Min. I can well believe it. But what *can* you do for me? Are their *Doppelgänger* available?

Euc. I fear not. The best thing I can do is to send you the Phantasm of a German Professor, a great friend of mine. He has read all books, and is ready to defend any thesis, true or untrue.

Min. A charming companion! And his name?

Euc. Phantasms have no names—only numbers. You may call him 'Herr Niemand,' or, if you prefer it, 'Number one-hundred-and-twenty-three-million-four hundred-and-fifty-six-thousand-seven-hundred-and-eighty-nine.'

Min. For *constant* use, I prefer 'Herr Niemand.' Let us now consider the question of the separation of Problems and Theorems.

§ 3. *The combination, or separation, of Problems and Theorems.*

Euc. I shall be glad to hear, first, the reasons given for separating them, and will then tell you *my* reasons for mixing them.

Min. I understand that the Committee of the Association for the Improvement of Geometrical Teaching, in their Report on the Syllabus of the Association, consider the separation as 'equivalent to the assertion of the principle that, while Problems are from their very nature dependent for the form, and even the possibility, of their solution on the arbitrary limitation of the instruments allowed to be used, Theorems, being truths involving no arbitrary element, ought to be exhibited in a form and sequence independent of such limitations.' They add however that 'it is probable that most teachers would prefer to introduce Problems, not as a separate section of

Geometry, but rather in connection with the Theorems with which they are essentially related.'

Euc. It seems rather a strange proposal, to print the Propositions in one order and read them in another. But a stronger objection to the proposal is that several of the Problems are Theorems as well—such as I. 46, for instance.

Min. How is that a Theorem?

Euc. It proves that there is such a thing as a Square. The definition, of course, does not assert real existence: it is merely provisional. Now, if you omit I. 46, what right would you have, in I. 47, to say '*draw* a Square'? How would you know it to be possible?

Min. We could easily deduce that from I. 34.

Euc. No doubt a Theorem might be introduced for that purpose: but it would be very like the Problem: you would have to say '*if* a figure were drawn under such and such conditions, it would be a Square.' Is it not quite as simple to draw it?

Then again take I. 31, where it is required to draw a Parallel. Although it has been proved in I. 27 that such things as parallel Lines *exist*, that does not tell us that, for every Line and for every point without that Line, there exists a real Line, parallel to the given Line *and passing through the given point.* And yet that is a fact essential to the proof of I. 32.

Min. I must allow that I. 31 and I. 46 have a good claim to be retained in their places: and if two are to be retained, we may as well retain all.

Euc. Another argument, for retaining the Problems

where they are, is the importance of keeping the numbering unchanged—a matter we have already discussed.

But perhaps the strongest argument is that it saves you from 'hypothetical constructions,' the danger of which has been so clearly pointed out by Mr. Todhunter (see pp. 222, 241).

Min. I think you have proved your case very satisfactorily. The next subject is 'the treatment of Pairs of Lines.' Would it not be well, before entering on this enquiry, to tabulate the Propositions that have been enunciated, whether as Axioms or Theorems, respecting them?

Euc. That will be an excellent plan. It will both give you a clear view of the field of your enquiry, and enable you to recognise at once any doubtful Axioms which you may meet with.

Min. Will you then favour me with your views on this matter?

Euc. Willingly. It is a subject which I need hardly say I considered very carefully before deciding what Definitions and Axioms to adopt.

---

§ 4. *Syllabus of propositions relating to Pairs of Lines.*

Let us begin with the simplest possible case, a Pair of infinite Lines which have two common points, and which therefore coincide wholly, and let us consider how such a Pair may be defined, and what other properties it possesses.

After that we will take a Pair of Lines which have a common point and a separate point ('a separate point' being one that lies on one of the Lines but not on the other), and which therefore have no other common point, and treat it in the same way.

And in the third place we will take a Pair of Lines which have *no* common point.

And let us understand, by 'the distance between two points,' the length of the right Line joining them; and, by 'the distance of a point from a Line,' the length of the perpendicular drawn, from the point, to the Line.

Now the properties of a Pair of Lines may be ranged under four headings:—

(1) as to common or separate points;

(2) as to the angles made with transversals;

(3) as to the equidistance, or otherwise, of points on the one from the other;

(4) as to direction.

We might distinguish the first two classes, which I have mentioned, as 'coincident' and 'intersecting': and these names would serve very well if we were going to consider only infinite Lines; but, as all the relations of infinite Lines, with regard to angles made with transversals, equidistance of points, and direction, are equally true of finite portions of them, it will be well to use names which will include them also. And the names I would suggest are 'coincidental' 'intersectional,' and 'separational.'

By 'coincidental Lines,' then, I shall mean Lines which either coincide or would do so if produced: and by 'intersectional Lines' I shall mean Lines which either intersect or would do so if produced; and, by 'separational Lines,' Lines which have no common point, however far produced.

In the same way, when I speak of 'Lines having a common point,' or of 'Lines having two common points,' I shall mean Lines which either have such points or would have them if produced.

It will also save time and trouble to agree on the use of a certain conventional phrase respecting transversals.

It admits of easy proof that, if a Pair of Lines make, with a certain transversal, either (*a*) a pair of alternate angles equal, or (*b*) an exterior angle equal to the interior opposite angle on the same side of the transversal, or (*c*) a pair of interior angles on the same side of the transversal supplementary; they will make, with the same transversal, (*d*) each pair of alternate angles equal, and (*e*) every exterior angle equal to the interior opposite angle on the same side of the transversal, and (*f*) each pair of interior angles on the same side of the transversal supplementary.

You will accept that as a simple Theorem, though with a somewhat lengthy enunciation?

Min. Certainly.

Euc. The phrase I propose is as follows. When I speak of a Pair of Lines as 'equally inclined to' a transversal, I wish it to be understood that they fulfil some one of the three conditions (*a*), (*b*), (*c*), and *therefore* all the three conditions (*d*), (*e*), (*f*).

Min. A most convenient abridgment.

Euc. Similarly, it admits of easy proof that, if a Pair of Lines make, with a certain transversal, either (*a*) a pair of alternate angles unequal, or (*b*) an exterior angle unequal to the interior opposite angle on the same side of the transversal, or (*c*) a pair of interior angles on the same side of the transversal not supplementary; they will make, with the same transversal, (*d*) each pair of alternate angles unequal, and (*e*) every exterior angle unequal to the interior opposite angle on the same side of the transversal, and (*f*) each pair of interior angles on the same side of the transversal not supplementary.

And when I speak of a Pair of Lines as 'unequally inclined to' a transversal, I wish it to be understood that they fulfil some one of the three conditions (*a*), (*b*), (*c*), and *therefore* all the three conditions (*d*), (*e*), (*f*).

Min. Very well.

Euc. Now the Propositions relating to Pairs of Lines may be divided into two classes, the first covering the ground occupied by my Axiom 10 ('two straight Lines cannot enclose a space') and my Propositions I. 16, 17, 27, 28, 31; the second that occupied by my Axiom 12 and Propositions I. 29, 30, 32. Those in the first class are logical deductions from Axioms which have never been disputed: the second class has furnished, through all ages, a battle-field for rival mathematicians. That *some one* of the Propositions in this class must be assumed as an Axiom is agreed on all hands, and each combatant in turn proclaims his own special favourite to be the *one* axiomatic truth of the series, insisting that all the rest ought to be proved as Theorems.

Let us now consider the properties of Pairs of Lines.

Such pairs may be arranged in three distinct classes. I will take them separately, and enumerate, for each class, first the 'subjects,' and secondly the 'predicates,' of Propositions concerning it.

Min. Let us make sure that we understand each other as to those two words. I presume that a 'subject' will include just so much 'property' as is needed to indicate the Pair of Lines referred to, i.e. to serve as a sufficient Definition for them?

Euc. Exactly so. Now, if we are told that a certain Pair of Lines fulfil some one of the following conditions:—

(1) they have two common points;

or (2) they have a common point, and are equally inclined to a certain transversal;

or (3) they have a common point, and one of them has two points on the same side of, and equidistant from, the other;

or (4) they have a common point and identical directions;

we may conclude that they fulfil *all* the following conditions:—

(1) they are coincidental;

(2) they are equally inclined to any transversal;

(3) they are 'equidistantial, i.e. any two points on one are equidistant from the other;

(4) they have identical directions.

Min. You mean, by 'conclude,' that we may *prove* our conclusion?

Euc. Yes, wherever proof is needed. Conclusions (1) and (4) need none, and are usually stated as Axioms.

Min. In subject (4), instead of 'identical directions,' why not say 'the same direction'?

Euc. Because I want to keep clearly in view that there are *two* Lines.

Min. In predicate (2), you speak of '*any* transversal': a little while ago, you spoke of '*every* exterior angle.' Do you make any distinction between 'any' and 'every'?

Euc. Where the things spoken of are limited in number, I use 'every'; where infinite, I use 'any' in order to bring the idea within the grasp of our finite intellects. For instance, you may talk of '*every* grain of sand in the world': there are, no doubt, what country-folk would call 'a good few' of them, but still the number is limited, and the mind can just grasp the idea. But if you tell me that '*every* cubic inch of Space contains eight cubic half-inches,' my mind is unable to form a distinct conception of the subject of your Proposition: you would convey the same truth, and in a form I *could* grasp, by saying '*any* cubic inch.'

Min. The angles made with the transversal are a little bewildering when the Pair of Lines shrinks, as it does in this case, into *one* Line. For instance, what becomes of the pair of interior angles on the same side of the transversal?

Euc. A diagram will make it clear.

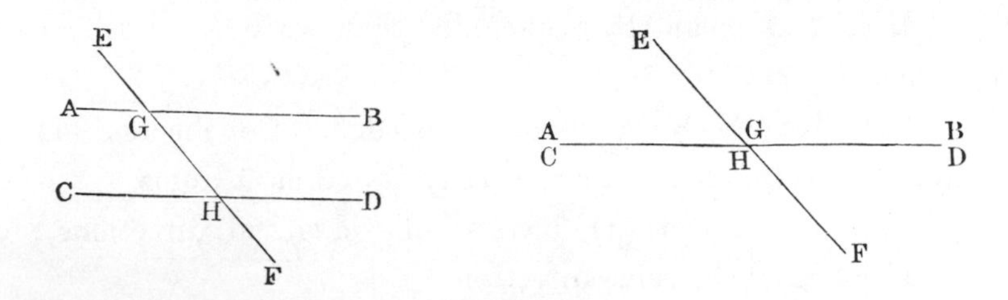

By examining the second figure (in which, as you see, there are three points with double names) we find that the alternate angles *AGF*, *EHD*, have become *vertical* angles; that the exterior and interior opposite angles *EGB*, *EHD*, have become *the same* angle; and that the two interior angles *BGF*, *DHE*, have become *adjacent* angles.

Min. That is quite clear.

Euc. Let us go on to the second class of Pairs of Lines.

If we are told that a certain Pair of Lines fulfil some one of the following conditions:—

(1) they have a common point and a separate point;

or (2) they have a common point, and are unequally inclined to a certain transversal;

or (3) they have a common point, and one of them has two points not-equidistant from the other;

or (4) they have a common point and different directions;

we may conclude that they fulfil *all* the following conditions:—

(1) they are separational;

(2) they are unequally inclined to any transversal;

(3) any two points on one, which are on the same side of the other, are not equidistant from it;

(4) a point may be found on each, whose distance from the other shall exceed any assigned length;

(5) they have different directions.

And thirdly, if we are told that a certain Pair of Lines fulfil some one of the following conditions:—

(1) they have a separate point, and are equally inclined to a certain transversal;

or (2) they have a separate point, and one of them has two points on the same side of, and equidistant from, the other;

we may conclude that they are separational.

Min. Why not use your own word 'parallel'?

Euc. Because that word is not uniformly employed, by modern writers, in one and the same sense. I would advise you, in discussing the works of my Modern Rivals, to disallow the use of the word 'parallel' altogether, and to oblige each writer to adopt a word which shall express his own definition.

Min. When you speak of two points on one Line, *which are on the same side of the other*, being 'equidistant from it,' do you include the case of their lying *on* the other Line?

Euc. Certainly. You may take them as lying on either side you like, and at zero-distances. The only case excluded is, where both points are *outside* the other Line, and on *opposite* sides of it.

Min. I understand you.

Euc. We shall find the Table of Propositions, which I now lay before you, very convenient to refer to. I have placed contranominal Propositions (i. e. Propositions of the form 'All *X* is *Y*,' 'All not-*Y* is not-*X*') in the same section.

TABLE I.

Containing twenty Propositions, of which some are undisputed Axioms, and the rest real and valid Theorems, deducible from undisputed Axioms.

[N.B. Those marked * have been proposed as Axioms.]

---

*1. A Pair of Lines, which have two common points, are coincidental.

or

*. Two Lines cannot enclose a space. [Euc. Ax.]

2. (*a*) A Pair of Lines, which have a separate point, have not two common points.

(*b*) A Pair of Lines, which have a common point and a separate point, are intersectional.

---

3. If there be given a Line and a point, it is possible to draw a Line, through the given point, intersectional with the given Line.

---

4. A Pair of intersectional Lines are unequally inclined to any transversal.

Cor. 1. In either pair of alternate angles, that, which is on the side, of the transversal, remote from the point of intersection, is the greater. [I. 16.]

Cor. 2. Every exterior angle, which is on the side, of the transversal, next to the point of intersection, is greater than the interior opposite angle on the same side. [I. 16.]

Cor. 3. The pair of interior angles, which are on the side, of the transversal, next to the point of intersection, are together less than two right angles. [I. 17.]

5. A Pair of Lines, which have a common point and are equally inclined to a certain transversal, are coincidental.

6. A Pair of Lines, which have a separate point and are equally inclined to a certain transversal, are separational. [I. 27, 28.]

---

7. If there be given a Line and a point without it, it is possible to draw a Line, through the given point, separational from the given Line. [I. 31.]

---

8. A Pair of intersectional Lines are such that any two points on one, which are on the same side of the other, are not equidistant from it.

Cor. That which is the more remote from the point of intersection has the greater distance.

9. A Pair of Lines, which have a common point and of which one has two points on the same side of and equidistant from the other, are coincidental.

10. A Pair of Lines, which have a separate point and of which one has two points on the same side of and equidistant from the other, are separational.

---

11. Each of a Pair of intersectional Lines has, in each portion of it, a point whose distance from the other exceeds any given length.

or

A Pair of intersectional Lines diverge without limit.

12. A Pair of Lines, which have two common points, have identical directions.

*13. (*a*) A Pair of Lines, which have different directions, have not two common points.

(*b*) A Pair of Lines, which have a common point and different directions, are intersectional.

---

*14. A Pair of intersectional Lines have different directions.

*15. A Pair of Lines, which have a common point and identical directions, are coincidental.

---

*16. If there be given a Line and a point without it: it is possible to draw a Line, through the given point, having a direction different from that of the given Line.

---

17. A Line, which has a point in common with one of two coincidental Lines has a point in common with the other also.

18. A Line, which has a point in common with one of two separational Lines, has a point separate from the other.

*19. A Line, which has a point in common with one of two separational Lines and also a point in common with the other, is intersectional with both.

---

*20. If there be three Lines; the first a right Line; the second, not assumed to be right, having a point separate from the first and being equidistantial from it; the third a right Line intersecting the first and diverging from it without limit on the side next to the second: the third is intersectional with the second.

Min. I see that 2 (*a*) is the contranominal of 1. But where does 2 (*b*) come from?

Euc. It is got from 2 (*a*) by adding, to each term, the property 'having a common point'—just as if we were to deduce, from 'all men are mortal,' 'all fat men are fat mortals.'

Min. You mean 5 to be contranominal to 4, I suppose. But 'coincidental' is not equivalent to 'non-intersectional.'

Euc. True: but I have added a new condition, viz. 'which have a common point,' to the subject. Non-intersectional Lines, which have a common point, are coincidental, just as, in the next Proposition, non-intersectional Lines, which have a separate point, are separational.

Min. 20 is rather a difficult enunciation to grasp.

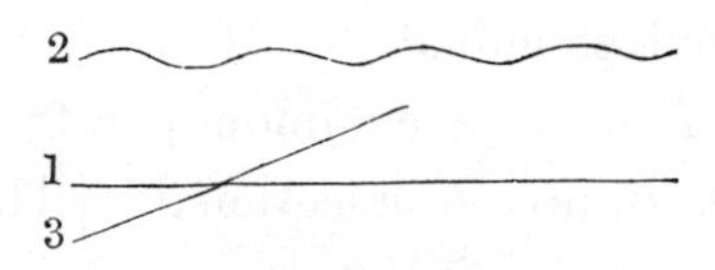

Euc. A diagram will make it clear. As a matter of fact, No. 2 would be a right Line: but, as we have no right, at present, to assume this, I have drawn it as a wavy line.

Min. I can suggest two Contranominals which you have omitted: one, deducible from 13 (*b*), 'Two Lines, which are not intersectional and which have different directions, have no common point, i.e. are separational'; the other, deducible from 15, 'Two Lines, which have a separate point and identical directions, have no common point, i.e. are separational.'

Euc. They are *valid* deductions, but in neither case do we know the 'subject' to be *real.*

Min. The 'contranominality'—if such a fearful word be allowable—of 17, 18, 19, seems obscure.

Euc. I will do what I can to make it less so.

Let us name the three Lines '*A*, *B*, *C*.'

Then 17 may be read 'A Line (*C*), which has a point in common with one (*A*) of two coincidental Lines (*A*, *B*), has a point in common with the other (*B*) also.'

From this we may deduce two Contranominals.

The first is 'If *A*, *C*, have a common point; and *B*, *C*, are separational: *A*, *B* have a separate point.' That is, 'a Line (*A*), which has a point in common with one (*C*) of two separational Lines (*B*, *C*), has a point separate from the other (*B*)': and thus we get 18.

The other Contranominal is 'If *A*, *C*, have a common point; and *A*, *B*, have a common point; and *B*, *C*, are separational: *A*, *B*, are intersectional.' That is, 'A Line (*A*), which has a point in common with one (*C*) of two separational Lines (*B*, *C*), and also a point in common with the other (*B*), is intersectional with that other (*B*).'

But we may evidently interchange *B* and *C* without interfering with the argument, and thus prove that *A* is *also* intersectional with *C*. Hence *A* is intersectional with *both*: and thus we get 19.

Min. That is quite clear.

Euc. We will now go a little further into the subject of separational Lines, as to which Table I. has furnished us with only three Propositions. There are, however, many other Propositions concerning them, which are fully admitted to be *true*, though no one of them has yet been proved from undisputed Axioms: and we shall find that they are so related to one another that, if any *one* be granted as an Axiom, all the rest may be proved; but, unless some one be so granted, none can be proved. Two thousand years of controversy have not yet settled the knotty question *which* of them, if any, can be taken as axiomatic.

If we are told that a certain Pair of Lines fulfil some one of the following conditions:—

(1) they are separational;

(2) they have a separate point and are equally inclined to a certain transversal;

(3) they have a separate point, and one of them has two points on the same side of and equidistant from the other;

we may prove (though not without the help of *some* disputed Axiom) that they fulfil *both* the following conditions:—

(1) they are equally inclined to any transversal;

(2) they are equidistantial from each other.

These Propositions, with the addition of my own I. 30, I. 32, and certain others, I will now arrange in a tabular form, placing Contranominals in the same section.

TABLE II.

Containing eighteen Propositions, of which no one is an undisputed Axiom, but all are real and valid Theorems, which, though not deducible from undisputed Axioms, are such that, if any one be admitted as an Axiom, the rest can be proved.

[N.B. Those marked * have been, or parts of them have been, proposed as Axioms.

---

1. A Pair of separational Lines are equally inclined to any transversal. [I. 29.]

*2. A Pair of Lines, which are unequally inclined to a certain transversal, are intersectional. [Euc. Ax.]

---

3. Through a given Point, without a given Line, a Line may be drawn such that the two Lines are equally inclined to any transversal.

---

4. A Pair of Lines, which are equally inclined to a certain transversal, are so to any transversal.

5. A Pair of Lines, which are unequally inclined to a certain transversal, are so to any transversal.

---

6. A Pair of separational Lines are equidistantial from each other.

*7. A Pair of Lines, of which one has two points on the same side of, and not equidistant from, the other, are intersectional.

*8. Through a given point, without a given Line, a Line may be drawn such that the two Lines are equidistantial from each other.

---

9. A Pair of Lines, of which one has two points on the same side of, and equidistant from, the other, are equally inclined to any transversal.

10. A Pair of Lines, which are unequally inclined to a certain transversal, are such that any two points on one, which are on the same side of the other, are not equidistant from it.

---

11. A Pair of Lines, which are equally inclined to a certain transversal, are equidistantial from each other.

12. A Pair of Lines, of which one has two points on the same side of, and not equidistant from, the other, are unequally inclined to any transversal.

---

13. A Pair of Lines, of which one has two points on the same side of, and equidistant from, the other, are equidistantial from each other.

14. A Pair of Lines, of which one has two points on the same side of, and not equidistant from, the other, are such that any two points on one, which are on the same side of the other, are not equidistant from it.

---

15. (*a*) A Pair of Lines, which are separational from a third Line, are not intersectional with each other.

(*b*) A Pair of Lines, which have a common point

and are separational from a third Line, are coincidental with each other.

or,

If there be given a Line and a point without it, only one Line can be drawn, through the given point, separational from the given Line.

(*c*) A Pair of Lines, which have a separate point and are separational from a third Line, are separational from each other. [I. 30.]

*16. (*a*) A Pair of intersectional Lines cannot both be separational from the same Line.

(*b*) A Line, which is intersectional with one of two separational Lines, is intersectional with the other also.

---

*17. A Line cannot recede from and then approach another; nor can one approach and then recede from another while on the same side of it.

---

18. (*a*) If a side of a Triangle be produced, the exterior angle is equal to each of the interior opposite angles. [I. 32.]

(*b*) The angles of a Triangle are together equal to two right angles. [I. 32.]

You will find it convenient to have the Propositions, that have been proposed as Axioms, repeated in a Table by themselves.

Table III.

Containing five Propositions, taken from Table II, which have been proposed as Axioms.

Euclid's Axiom.

A Pair of Lines, which have a separate point and make, with a certain transversal, two interior angles on one side of it together less than two right angles, are intersectional on that side.

[This is one case of II. 2, with an additional statement as to the *side* of the transversal on which the Lines will meet.]

T. Simpson's Axiom.

A Pair of Lines, which have a separate point and of which one has two points on the same side of, and not equidistant from, the other, are intersectional.

[*This is* II. 7.]

Clavius' Axiom.

Through a given Point, without a given Line, a Line may be drawn equidistantial from the given Line.

[*This is part of* II. 8.]

Playfair's Axiom.

A pair of intersectional Lines cannot both be separational from the same Line.

[*This is* II. 16 (*a*).]

R. Simpson's Axiom.

A Line cannot recede from and then approach another: nor can one approach and then recede from another while on the same side of it.

[*This is* II. 17.]

Min. In the predicate of 2, what right have you to say 'are intersectional'? The true contradictory of 'separational' would be 'have a common point.'

Euc. True: but we may assume as an Axiom 'A Pair of coincidental Lines are equally inclined to any transversal.' This, combined with 1, gives 'A Pair of not-intersectional Lines are equally inclined to any transversal,' whose Contranominal is 2.

Similarly, we may combine, with 6, the Axiom 'A Pair of coincidental Lines are equidistantial from each other,' and thus get a Theorem whose Contranominal is 7.

Min. In classing 15 (*a*), (*b*), and (*c*) under one number, you mean, I suppose, that they are so related that, if any one of them be granted, the others may be deduced?

Euc. Certainly.

Min. I see that if (*a*) be given, (*b*) may be deduced by simply adding 'having a common point' to subject and predicate. And I see that (*b*) and (*c*) are Contranominals, so that, if either be given, the other follows. But I don't see how, if (*b*) only were given, you would prove (*a*).

Euc. You can prove (*c*) from it, as you say: and then, from (*b*) and (*c*) combined, you can prove (*a*) thus:—

Any Pair of Lines, which are separational from a third Line, must belong to one or both of the two classes, 'having a common point,' 'having a separate point.' Hence if we take these two classes together, we include *any* Pair that can be proposed. Thus we get the Theorem '*Any* Pair of Lines, which are separational from a third Line, are either coincidental or separational'; the predicate of which is equivalent to 'are not intersectional.'

Min. I see. And how are 16 (*a*) and 16 (*b*) related to 15 ?

Euc. Each of them is a Contranominal of 15 (*a*); and they are also contranominal to each other.

Min. I should like to see that drawn out.

Euc. Let '*A*, *B*, *C*' be three Lines. Then 16 (*a*) may be written 'A Pair of Lines (*A*, *B*), which are separational from a third Line (*C*), are not intersectional with each other.'

This yields three Contranominals. The first is 'If *A*, *B*, are intersectional; it cannot be true that *B*, *C*, are separational, and also *A*, *C*.' i. e. 'A Pair of intersectional Lines (*A*, *B*) cannot both be separational from a third Line (*C*)': the second is 'If *B*, *C* are separational, and *A*, *B* intersectional; then *A*, *C* are not separational.' i. e. 'A Line (*A*) which is intersectional with one (*B*) of two separational Lines (*B*, *C*), is not separational from the other (*C*)': and the third proves a similar Theorem for *B*.

Min. Yes, but your conclusion *now* is '*A* is not separational from *C*': whereas 16 (*b*) says 'is intersectional.'

Euc. That is so: but since *A* is intersectional with a Line (*B*) which is separational from *C*, it is axiomatic that it has a point separate from *C*, and so cannot be coincidental with it. Hence, its being 'not separational from *C*' proves that it must be intersectional with it.

Min. I suppose I must take it on trust that any one of these 18 is sufficient logical basis for the other 17: I can hardly ask you to go through 306 demonstrations!

Euc. I can do it with 11. You will grant me that, when two Propositions are contranominal, so that each can be proved from the other, I may select either of the two for my series of proofs, but need not include *both*?

Min. Certainly.

Euc. Here are the proofs, which you can read afterwards at your leisure. (See Appendix III.)

---

§ 5. *Playfair's Axiom.*

Euc. The next *general* question to be discussed is the proposed substitution of Playfair's Axiom for mine. With regard to mine, I am quite ready to admit that it is not axiomatic until Prop. 17 has been proved. What is an Axiom at one stage of our knowledge is often anything but an Axiom at an earlier stage.

Min. The great question is whether it is axiomatic *then*.

Euc. I am quite aware of that: and it is because this is not only the great question of the whole First Book, but also the crucial test by which my method, as compared with those of my 'Modern Rivals,' must stand or fall, that I entreat your patience in speaking of a matter which cannot possibly be dismissed in a few words.

Min. Pray speak at whatever length you think necessary to so vital a point.

Euc. Let me remark in the first place—it is a minor matter, but yet one that *must* come in somewhere, and I do not want to break the thread of my argument—that we need, in any complete geometrical treatise, *some* practical geometrical test by which we can prove that two given finite Lines will meet if produced. My Axiom serves this purpose—a secondary purpose it is true—but it is incumbent on any one, who proposes to do away with it, to provide some sufficient substitute.

Min. I admit all that.

Euc. Now, if the test I propose—that the two Lines make with a certain transversal two interior angles on the same side of it together less than two right angles—be objected to as not sufficiently simple, the question arises, what simpler test can be proposed?

Min. The supporters of Playfair's Axiom would of course reply 'that one of the two Lines should cut a Line known to be parallel to the other.'

Euc. Assuming that what is needed is a distinct conception of the geometrical relationship of the two Lines, whose future meeting we are asked to believe in, which picture, think you, is the more likely to yield us such a conception—two finite Lines, both intersected by a transversal, and having a known angular relation to that transversal and so to each other—or two Lines 'known to be parallel,' that is two Lines of whose geometrical relationship, so far as our field of vision extends, we know absolutely nothing, but can only say that, in the far-away region of infinity, they do *not* meet?

Min. In clearness of conception, your picture seems to

have the advantage. In fact, I could not form any mental picture *at all* of the relative position of two finite Lines, if *all* I knew about them was their never meeting however far produced: and it would be equally impossible to form any mental picture of the position which a Line, crossing one of them, would have relatively to the other. But, though your picture may be *more* easy to conceive, I doubt if it is enough so to constitute an axiom.

Euc. Taken by itself, it may be, as you say, not entirely axiomatic. But I think I can put before you a few considerations which will make it more acceptable.

Min. They will be well worth having. An absolute *proof* of it, from first principles, would be received, I can assure you, with absolute *rapture*, being an *ignis fatuus* that mathematicians have been chasing from your age down to our own.

Euc. I know it. But I cannot help you. Some mysterious flaw lies at the root of the subject. Probabilities are all I have to offer you.

Now suppose you were assured, with regard to two finite Lines placed before you, that, when produced in a certain direction, one of them *approached* the other, that is, contained two points, of which the second was nearer, to the other Line, than the first, would you not think it probable—if not absolutely certain—that they would meet at last?

Min. Utilising—as I suppose you will allow me to do—my knowledge of the properties of *asymptotes*, I should say 'No. The mere fact of *approach*, granted as to two Lines, does *not* secure a future *meeting*.'

Euc. But, if you look into the depths of your own consciousness—assuming such depths to exist—you will find, I believe, an eternal distinction maintained, in this respect, between straight and curved Lines: so that Lines of the one kind *must*, if they approach, ultimately meet, whereas those of the other kind need not.

Min. I will grant it, provisionally, if only to know what you are going to deduce from it.

Euc. I will now ask you to consider this diagram.

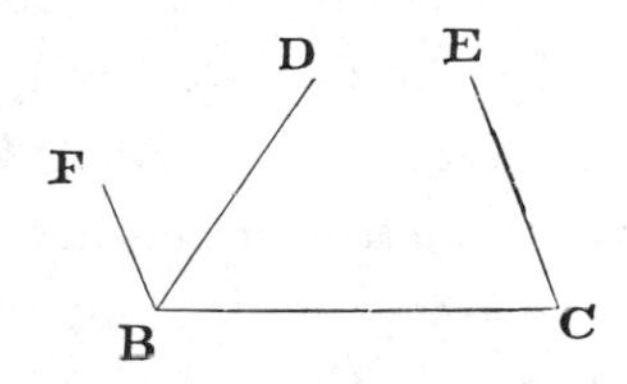

Suppose it given that the Lines *BD, CE,* make with *BC* two angles together less than two right angles. My object is to show that probably—if not certainly—they will meet, if produced towards *D*, *E*.

Let *BF* be so drawn that the angles *FBC, BCE*, may be together equal to two right angles.

Now, if any point in *BD* be nearer to *CE* than *B* is, what is required is proved, since *BD approaches CE*.

But, if this be not so, then *F* (which is obviously further from *CE* than some point in *BD* is) must also be further from *CE* than *B* is; i.e. *FB* must *approach EC*; i.e. *FB* and *EC* must ultimately meet, below *BC*, and so form a Triangle, whose angles at *B* and *C* will be (by my Prop. 17) less than two right angles. Hence the angles *FBC, BCE*, must be *greater* than two right angles, since the four angles are (by my Prop. 13) equal to four right angles. But this is absurd, since they were made *equal* to two right angles.

Hence *D is* nearer to *CE* than *B* is; i.e. *BD approaches CE*, and so will meet it if produced.

Min. You certainly *have* made your Axiom a little more axiomatic. It is, I presume, an afterthought of yours: otherwise you would have made your Axiom deal with *approaching* Lines, and would then have proved your present Axiom as a Theorem.

Euc. Excuse me. Whatever the habits of modern geometricians may be, in *our* day we always investigated a subject down to the very roots. No 'afterthought' was possible. You of the nineteenth century may 'look before and after,' if it so please you, so long as *we* have liberty to look at what is at our feet: *you* may 'sigh for what is not,' and welcome, so long as *we* may chuckle at what *is*!

Min. Flippancy will not serve your turn. If you have no better reason than *that*—

Euc. I *have* a better reason. How could I have dealt with *approaching* Lines without first strictly defining 'the distance of a point from a Line'?

Min. Nohow, I grant you.

Euc. Which would have entailed a definition of 'the distance of a point from a point,' i.e. the length of the *shortest* path by which the one can pass to the other—which again would have entailed the comparison of all possible paths—which again would have entailed the estimation of the lengths of curved Lines—which again—

Min. This is uncanny! It is whichcraft!

Euc. (*preserves a disgusted silence*).

Min. I beg your pardon. I grant that you have made out a very good case for your own Axiom, and but a bad one for Playfair's.

Euc. I will make it worse yet, before I have done. My next remark will be best explained with the help of a diagram.

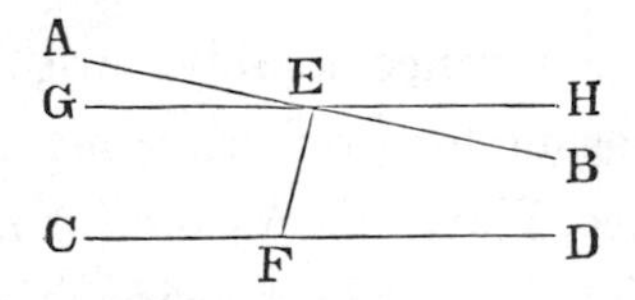

Let *AB* and *CD* make, with *EF*, the two interior angles *BEF*, *EFD*, together less than two right angles. Now if through *E* we draw the Line *GH* such that the angles *HEF*, *EFD* may be equal to two right angles, it is easy to show (by Prop. 28) that *GH* and *CD* are 'separational.'

Min. Certainly.

Euc. We see, then, that any Lines which have the property (let us call it 'a') of making, with a certain transversal, two interior angles together less than two right angles, have also the property (let us call it 'β') that one of them intersects a Line which is separational from the other.

Min. I grant it.

Euc. Now suppose you decline to grant my 12th Axiom, but are ready to grant Playfair's Axiom, that two intersectional Lines cannot both be separational from the same Line: then you have in fact granted my Axiom.

Min. Be good enough to prove that.

Euc. Lines, which have property 'a,' have property

'β.' Lines, which have property 'β,' meet if produced; for, if not, there would be two Lines both separational from the same Line, which is absurd. Hence Lines, which have property 'a,' meet if produced.

Min. I see now that those who grant Playfair's Axiom have no right to object to yours: and yours is certainly the more simple one.

Euc. To make assurance doubly sure, let me give you two additional reasons for preferring my Axiom.

In the first place, Playfair's Axiom (or rather the Contranominal of it which I have been using, that 'a Line which intersects one of two separational Lines will also meet the other') does not tell us *which way* we are to expect the Lines to meet. But this is a very important matter in constructing a diagram.

Min. We might obviate that objection by re-wording it thus:—'If a Line intersect one of two separational Lines, that portion of it which falls between them will, if produced, meet the other.'

Euc. We might: and therefore I lay little stress on *that* objection.

Euc. In the second place, Playfair's Axiom asserts *more* than mine does: and all the additional assertion is superfluous, and a needless strain on the faith of the learner.

Min. I do not see that in the least.

Euc. It *is* rather an obscure point, but I think I can make it clear. We know that all Pairs of Lines, which have property 'a,' have also property 'β'; but we do *not* know as yet (till we have proved I. 29) that all, which have property 'β,' have also property 'a.'

Min. That is so.

Euc. Then, for anything we know to the contrary, class 'β' *may* be larger than class 'a.' Hence, if you assert anything of class 'β,' the logical effect is more extensive than if you assert it of class 'a': for you assert it, not only of that portion of class 'β' which is known to be included in class 'a,' but also of the unknown (but possibly existing) portion which is *not* so included.

Min. I see that now, and consider it a real and very strong reason for preferring your axiom.

But so far you have only answered Playfair. What do you say to the objection raised by Mr. Potts? 'A stronger objection appears to be that the converse of it forms Euc. I. 17; for both the assumed Axiom and its converse should be so obvious as not to require formal demonstration.'

Euc. Why, I say that I deny the general law which he lays down. (It is, of course, the *technical* converse that he means, not the *logical* one. 'All X is Y' has for its technical converse 'All Y is X'; for its logical, 'Some Y is X.') Let him try his law on the Axiom 'All right angles are equal,' and its technical converse 'All equal angles are right'!

Min. I withdraw the objection.

---

§ 6. *The Principle of Superposition.*

Min. The next subject is the principle of 'superposition.' You use it twice only (in Props. 4 and 8) in the First

Book: but the modern fancy is to use it on all possible occasions. The Syllabus indicates (to use the words of the Committee) 'the free use of this principle as desirable in many cases where Euclid prefers to keep it out of sight.'

Euc. Give me an instance of this modern method.

Min. It is proposed to prove I. 5 by taking up the isosceles Triangle, turning it over, and then laying it down again *upon itself*.

Euc. Surely that has too much of the Irish Bull about it, and reminds one a little too vividly of the man who walked down his own throat, to deserve a place in a strictly philosophical treatise?

Min. I suppose its defenders would say that it is conceived to leave a trace of itself behind, and that the reversed Triangle is laid down upon the trace so left.

Euc. That is, in fact, the same thing as conceiving that there are *two* coincident Triangles, and that one of them is taken up, turned over, and laid down upon the other. And what does their subsequent coincidence prove? Merely this: that the right-hand angle of the first is equal to the left-hand angle of the second, and *vice versâ*. To make the proof complete, it is necessary to point out that, owing to the original coincidence of the Triangles, this same 'left-hand angle of the second' is *also* equal to the *left*-hand angle of the first: and then, and not till then, we may conclude that the base-angles of the first Triangle are equal. This is the full argument, strictly drawn out. The Modern books on Geometry often attain their much-vaunted brevity by the dangerous process of omitting links in the chain; and some of the new proofs, which at first sight seem to be shorter than mine, are really longer when fully stated. In this particular case I think you will allow that I had good reason for not adopting the method of superposition?

Min. You had indeed.

Euc. Mind, I do not object to that proof, if appended to mine as an *alternative*. It will do very well for more advanced students. But, for beginners, I think it much clearer to have two non-isosceles Triangles to deal with.

Min. But your objection to laying a Triangle down *upon itself* does not apply to such a case as I. 24.

Euc. It does not. Let us discuss that case also. The Moderns would, I suppose, take up the Triangle *ABC*, and apply it to *DEF* so that *AB* should coincide with *DE*?

Min. Yes.

Euc. Well, that would oblige you to say 'and join *C*, in its new position, to *E* and *F*.' The words 'in its new position' would be necessary, because you would now have *two* points in your diagram, both called '*C*.' And you would also be obliged to give the points *D* and *E* additional names, namely '*A*' and '*B*.' All which would be very confusing for a beginner. You will allow, I think, that I was right here in constructing a new Triangle instead of transferring the old one?

Min. Cuthbertson evades that difficulty by re-naming the point *C*, and calling it '*Q*.'

Euc. And do the points *A* and *B* take their names with them?

Min. No. They adopt the names '*D*' and '*E.*'

Euc. It is very like making a new Triangle!

Min. It is indeed. I think you have quite disposed of the claims of 'superposition.' The only remaining subject for discussion is the omission of the diagonals in Book II.

---

§ 7. *The omission of diagonals in Euc. II.*

Euc. Let us test it on my II. 4. We will go through *my* proof of it, and then the proof given by some writer who ignores the diagonal, supplying if necessary any of those gaps in argument which my Modern Rivals so often indulge in, and which give to their proofs a delusive air of neatness and brevity.

'*If a Line be divided into any two parts, the square of the Line is equal to the squares of the two parts with twice their rectangle.*

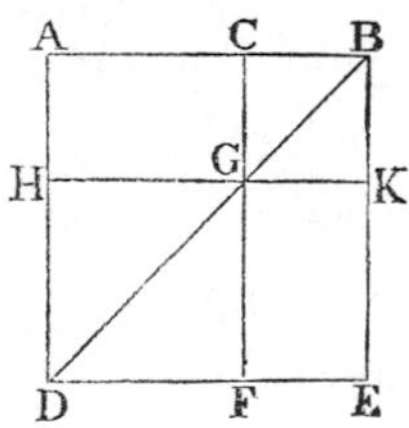

Let AB be divided at C. It is to be proved that square of AB is equal to squares of AC, CB, with twice rectangle of AC, CB.

On AB describe Square $ADEB$; join BD; from C

draw CF parallel to AD or BE, cutting BD at G; and through G draw HK parallel to AB or DE.

∵ BD cuts Parallels AD, CF,

∴ exterior angle CGB = interior opposite angle ADB. [I. 29

also ∵ $AD = AB$,

∴ angle ADB = angle ABD; [I. 5

∴ angle CGB = angle ABD;

∴ $CG = CB$; [I. 6

but $BK = CG$, and $GK = CB$; [I. 34

∴ CK is equilateral.

also, ∵ angle CBK is right,

∵ CK is rectangular; [I. 46. Cor.

∴ CK is a Square.

Similarly HF is a Square and = square of AC, for $HG = AC$. [I. 34

Also, ∵ AG, GE are equal, being complements, [I. 43

∴ AG and GE = twice AG;

= twice rectangle of AC, CB.

But these four figures make up AE.

Therefore the square of AB &c. Q. E. D.'

That is just 128 words, counting from 'On AB describe' down to the words 'rectangle of AC, CB.' What author shall we turn to for a rival proof?

Min. I think Wilson will be best.

Euc. Very well. Do the best you can for him. You may use all my references if you like, and if you can do so legitimately.

Min. 'Describe Square $ADEB$ on AB. Through C draw CF parallel to AD, meeting——'

Euc. You must insert 'or BE,' to make the comparison fair.

Min. Certainly. I will mark the necessary insertions by parentheses. 'Through C draw CF parallel to AD (or BE), meeting DE in F.'

Euc. You may omit those four words, as they do not occur in my proof.

Min. Very well. 'Cut off (from CF) $CG = CB$. Through G draw HK parallel to AB (or DE). It is easily shewn that CK, HF are squares of CB, AC; and that AG, GE, are each of them rectangle of AC, CB.'

Euc. We can't admit 'it is easily shewn'! He is bound to *give* the proof.

Min. I will do it for him as briefly as I can. '$\because CG = CB$, and $BK = CG$, and $GK = CB$, $\therefore CK$ is equilateral. It is also rectangular, since angle CBK is right. $\therefore CK$ is a Square.' I'm afraid I mustn't say 'Similarly HF is a Square'?

Euc. Certainly not: it requires a different proof.

Min. 'Because $CF = AD = AB$, and CG, CB, parts of them, are equal, $\therefore$ remainder $GF =$ remainder $AC, = HG$. But $HD = GF$, and $DF = HG$; $\therefore HF$ is equilateral. It is also rectangular, since angle HDF is right. $\therefore HF$ is a Square, and = square of AC. Also AG is rectangle of AC, CB.' I fear I can't assume GE to be equal to AG?

Euc. I fear I cannot permit you to assume the truth of my I. 43.

Min. 'Also GE is rectangle of AC, CB, since $GF = AC$, and $GK = CB$. $\therefore AG$ and $GE =$ twice rectangle of AC, CB.'

Euc. That will do. How many words do you make it?

Min. 145.

Euc. Then the omission of the diagonal, instead of shortening the proof, has really lengthened it by seventeen words! Well! Has it any advantage in the way of neatness to atone for its greater length?

Min. Certainly not. It is quite unsymmetrical. I very much prefer your method of appealing to the beautiful Theorem of the equality of complements.

Euc. Then that concludes our present interview: we will meet again when you have reviewed my Modern Rivals one by one. If you had any slow music handy, I would vanish to it: as it is——

(*Vanishes without slow music.*)

ACT II.

Manuals which reject Euclid's treatment of Parallels.

SCENE I.

'E fumo dare lucem.

[MINOS *sleeping. To him enter, first a cloud of tobacco-smoke; secondly the bowl, and thirdly the stem, of a gigantic meerschaum; fourthly the phantasm of* HERR NIEMAND, *carrying a pile of phantom-books, the works of Euclid's Modern Rivals, phantastically bound.*]

Niemand. The first author we have to consider is M. Legendre, is it not?

Minos. (*aside*) Not a single word of greeting! He plunges *in medias res* with a more fearful suddenness than Euclid himself! (*Aloud*) It is so, mein lieber Herr.

Nie. No time to waste in civil speeches! It is for you to question, for me to answer. I have read M. Legendre's book. Ach! It is beautiful! You shall find in it no flaw!

Min. I do not expect to do so.

ACT II.

SCENE II.

Treatment of Parallels by methods involving infinite series.

LEGENDRE.

'Fine by degrees, and beautifully less.'

Nie. I lay before you '*Éléments de Géométrie*' by Mons. A. M. LEGENDRE, the 14th edition, 1860.

Min. Let me begin by asking you (since I consider you and your client as one in this matter) how you define a straight Line.

Nie. As 'the shortest path from one point to another.'

Min. This does not seem to me to embody the primary idea which the word 'straight' raises in the mind. Is not the natural process of thought to realise *first* the notion of 'a straight Line,' and *then* to grasp the fact that it is the shortest path between two points?

Nie. That may be the natural process: but surely you will allow our Definition to be a legitimate one?

Min. I think not: and I have the great authority of Kant to support me. In his 'Critique of Pure Reason,' he says (I quote from Meiklejohn's translation, in Bohn's Philosophical Library, pp. 9, 10), 'Mathematical judgments are always synthetical . . . "A straight Line between two points is the shortest" is a synthetical Proposition. For my conception of *straight* contains no notion of *quantity*, but is merely *qualitative.* The conception of the *shortest* is therefore wholly an addition, and by no analysis can it be extracted from our conception of a straight Line.'

This may fairly be taken as a denial of the fitness of the Axiom to stand as a Definition. For all Definitions ought to be the expressions of analytical, not of synthetical, judgments: their predicates ought not to introduce anything which is not already included in the idea corresponding to the subject. Thus, if the idea of 'shortest distance' cannot be obtained by a mere analysis of the conception represented by 'straight Line,' the Axiom ought not to be used as a Definition.

Nie. We are not particular as to whether it be taken as a Definition or Axiom: either will answer our purpose.

Min. Let us then at least banish it from the *Definitions.* And now for its claim to be regarded as an *Axiom.* It involves the assertion that a straight Line is shorter than any *curved* Line between the two points. Now the length of a curved Line is altogether too difficult a subject for a beginner to have to consider: it is moreover unnecessary that he should consider it at all, at least in the earlier parts of Geometry: all he really needs is to grasp the fact that it is shorter than any *broken* Line made up of straight Lines.

Nie. That is true.

Min. And all cases of broken Lines may be deduced from their simplest case, which is Euclid's I. 20.

Nie. Well, we will abate our claim and simply ask to have I. 20 granted us as an Axiom.

Min. But it can be *proved* from your own Axioms: and it is a generally admitted principle that, at least in dealing with beginners, we ought not to take as axiomatic any Theorem which can be proved by the Axioms we already possess.

Nie. For *beginners* we must admit that Euclid's method of treating this point is the best. But you will allow ours to be a legitimate and elegant method for the advanced student?

Min. Most certainly. The whole of your beautiful treatise is admirably fitted for advanced students: it is only from the *beginner's* point of view that I venture to criticise it at all.

Your treatment of angles and right angles does not, I think, differ much from Euclid's?

Nie. Not much. We *prove*, instead of assuming, that all right angles are equal, deducing it from the Axiom that two right Lines cannot enclose a space.

Min. I think some such proof a desirable interpolation. I will now ask you how you prove Euc. I. 29.

Nie. What preliminary Propositions will you grant us as proved?

Min. Euclid's series consists of Ax. 12, Props. 4, 5, 7, 8, 13, 15, 16, 27, 28. I will grant you as much of that series as you have proved by methods not radically differing from his.

Nie. That is, you grant us Props. 4, 13, and 15. Prop. 16 is not in our treatise. The next we require is Prop. 6.

Min. That you may take as proved.

Nie. And, next to that, Prop. 20: *that* we assume as an Axiom, and from it, with the help of Prop. 6, we deduce Prop. 19.

Min. For our present purpose you may take Prop. 19 as proved.

Nie. From Props. 13 and 19 we deduce Prop. 32; and from that, Ax. 12; from which Prop. 29 follows at once.

Min. Your proof of Prop. 32 is long, but beautiful. I need not, however, enter on a discussion of its merits. It is enough to say that what we require is a proof suited to the capacities of *beginners*, and that this Theorem of yours (Prop. XIX, at p. 20) contains an infinite series of Triangles, an infinite series of angles, the terms of which continually decrease so as to be ultimately less than any assigned angle, and magnitudes which vanish simultaneously. These considerations seem to me to settle the question. I fear that your proof of this Theorem, though a model of elegance and perspicuity as a study for the advanced student, is wholly unsuited to the requirements of a beginner.

Nie. That we are not prepared to dispute.

Min. It seems superfluous, after saying this, to ask what test for the meeting of Lines you have provided:

but we may as well have that stated, to complete the enquiry.

Nie. We give Euclid's 12th Axiom, which we prove from Prop. 32, using the principle of Euc. X. 1 (second part), that 'if the greater of two unequal magnitudes be bisected, and if its half be bisected, and so on; a magnitude will at length be reached less than the lesser of the two magnitudes.'

Min. That again is a mode of proof entirely unsuited to beginners.

The general style of your admirable treatise I shall not attempt to discuss: it is one I would far rather take as a model to imitate than as a subject to criticise.

I can only repeat, in conclusion, what I have already said, that your book, though well suited for advanced students, is not so for beginners.

Nie. At this rate we shall make short work of the twelve Modern Rivals!

ACT II.

SCENE III.

Treatment of Parallels by angles made with transversals.

COOLEY.

'The verbal solemnity of a hollow logic.'
COOLEY, *Pref.* p. 20.

Nie. I have now the honour to lay before you '*The Elements of Geometry, simplified and explained,*' by W. D. COOLEY, A.B., published in 1860.

Min. Please to hand me the book for a moment. I wish to read you a few passages from the Preface. It is always satisfactory—is it not?—to know that a writer, who attempts to 'simplify' Euclid, begins his task in a becoming spirit of humility, and with some reverence for a name that the world has accepted as an authority for two thousand years.

Nie. Truly.

MINOS *reads*.

'The Elements of Plane Geometry . . . are here presented in the reduced compass of 36 Propositions, perfectly coherent, fully demonstrated, and reaching quite as far as the 173 Propositions contained in the first six books of Euclid.' Modest, is it not?

Nie. A little high-flown, perhaps. Still, you know, if they really *are* 'fully demonstrated'——

Min. If! In page 4 of the Preface he talks of 'Euclid's circumlocutory shifts': in the same page he tells us that 'the doctrine of proportion, as propounded by Euclid, runs into prolixity though wanting in clearness': and again, in the same page, he states that most of Euclid's *ex absurdo* proofs 'though containing little,' yet 'generally puzzle the young student, who can hardly comprehend why gratuitous absurdities should be so formally and solemnly dealt with. These Propositions therefore are omitted from our Book of Elements, and the Problems also, for the science of Geometry lies wholly in the Theorems. Thus simplified and freed from obstructions, the truths of Geometry may, it is hoped, be easily learned, even by the youngest.' But perhaps the grandest sentence is at the end of the Preface. 'Then as to those Propositions (the first and last of the 6th Book), in which, according to the same authority' (he is alluding to the Manual of Euclid by Galbraith and Haughton), 'Euclid so beautifully illustrates his celebrated Definition, they appear to our eyes to exhibit only the verbal solemnity of a hollow logic, and to exemplify nothing but the formal application of a nugatory principle.' Now let us see, mein Herr, whether Mr. Cooley has done anything worthy of the writer of such 'brave 'orts' (as Shakespeare has it): and first let me ask how you define Parallel Lines.

NIEMAND *reads.*

'Right Lines are said to be parallel when they are equally and similarly inclined to the same right Line, or make equal angles with it towards the same side.'

Min. That is to say, if we see a Pair of Lines cut by a certain transversal, and are told that they make equal angles with it, we say 'these Lines are parallel'; and conversely, if we are told that a Pair of Lines are parallel, we say 'then there *is* a transversal, *somewhere*, which makes equal angles with them'?

Nie. Surely, surely.

Min. But we have no means of finding it? We have no right to draw a transversal at random and say '*this* is the one which makes equal angles with the Pair'?

Nie. Ahem! Ahem! Ahem!

Min. You seem to have a bad cough.

Nie. Let us go to the next subject.

Min. Not till you have answered my question. *Have* we any means of finding the particular transversal which makes the equal angles?

Nie. I am sorry for my client, but, since you are so *exigeant*, I fear I must confess that we have *no* means of finding it.

Min. Now for your proof of Euc. I. 32.

Nie. You will allow us a preliminary Theorem?

Min. As many as you like.

Nie. Well, here is our Theorem II. '*When two parallel straight Lines AB, CD, are cut by a third straight Line EF, they make with it the alternate angles AGH, GHD, equal; and also the two internal angles at the same side BGH, GHD equal to two right angles.*

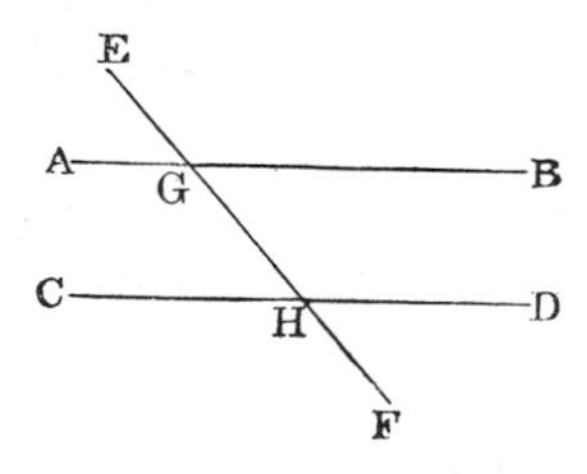

For *AGH* and *EGB* are equal because vertically opposite, and *EGB* is also equal to *GHD* (Definition); therefore—'

Min. There I must interrupt you. How do you know that *EGB* is equal to *GHD*? I grant you that, by the Definition, *AB* and *CD* make equal angles with *a certain* transversal: but have you any ground for saying that *EF* *is* the transversal in question?

Nie. We have not. We surrender at discretion. You will permit us to march out with the honours of war?

Min. We grant it you of our royal grace. March him off the table, and bring on the next Rival.

ACT II.

SCENE IV.

Treatment of Parallels by equidistances.

CUTHBERTSON.

'Thou art so near, and yet so far.'
Modern Song.

Nie. I now lay before you '*Euclidian Geometry,*' by FRANCIS CUTHBERTSON, M.A., late Fellow of C. C. C., Cambridge; Head Mathematical Master of the City of London School; published in 1874.

Min. It will not be necessary to discuss with you *all* the innovations of Mr. Cuthbertson's book. The questions of the separation of Problems and Theorems, the use of superposition, and the omission of the diagonals in Book II, are general questions which I have considered by themselves. The only points, which you and I need consider, are the methods adopted in treating Right Lines, Angles, and Parallels, wherever those methods differ from Euclid's.

The first subject, then, is the Right Line. How do you define and test it?

Nie. As in Euclid. But we *prove* what Euclid has assumed as an Axiom, namely, that two right Lines cannot have a common segment.

Min. I am glad to hear you assert that Euclid has assumed it 'as an Axiom,' for the interpolated and illogical corollary to Euc. I. 11 has caused many to overlook the fact that he has assumed it as early as Prop. 4, if not in Prop. 1. What is your proof?

NIEMAND *reads.*

'*Two straight Lines cannot have a common segment.*'

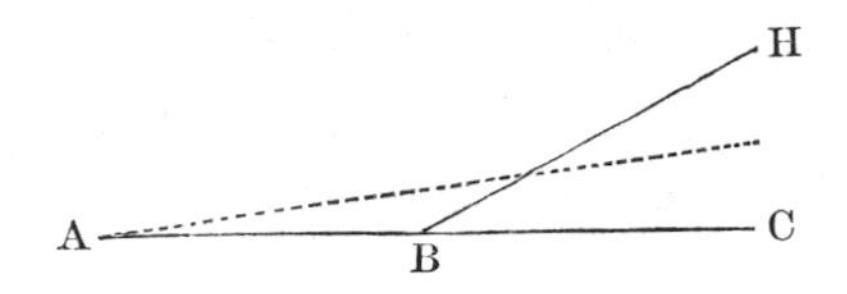

'For if two straight Lines *ABC*, *ABH* could have a common segment *AB*; then the straight Line *ABC* might be turned about its extremity *A*, towards the side on which *BH* is, so as to cut *BH*; and thus two straight Lines would enclose a space, which is impossible.'

Min. You assume that, *before C* crosses *BH*, the portions coinciding along *AB* will diverge. But, if *ABH* is a right Line, this will not happen till *C* has *passed H*.

Nie. But you would then have one portion of the revolving Line in motion, and another portion at rest.

Min. Well, why not?

Nie. We may assume *that* to be impossible; and that,

if a Line revolves about its extremity, it all moves at once.

Min. Which, I take the liberty to think, is *quite* as great an assumption as Euclid's. I think the Axiom quite plain enough without any proof.

Your treatment of angles, and right angles, is the same as Euclid's, I think?

Nie. Yes, except that we *prove* that 'all right angles are equal.'

Min. Well, it *is* capable of proof, and therefore had better not be retained as an Axiom.

I must now ask you to give me your proof of Euc. I. 32.

Nie. We prove as far as I. 28 as in Euclid. In order to prove I. 29, we first prove, as a Corollary to Euc. I. 20, that 'the shortest distance between two points is a straight Line.'

Min. What is your next step?

Nie. A Problem (Pr. F. p. 52) in which we prove the Theorem that, of all right Lines drawn from a point to a Line, the perpendicular is the least.

Min. We will take that as proved.

Nie. We then deduce that the perpendicular is the shortest path from a point to a Line.

Next comes a Definition. 'By the distance of a point from a straight Line is meant the shortest path from the point to the Line.'

Min. Have you anywhere defined the distance of one point from another?

Nie. No.

Min. We had better have that first.

Nie. Very well. 'The distance of one point from another is the shortest path from one to the other.'

Min. Might we not say 'is the length of the right Line joining them?'

Nie. Yes, that is the same thing.

Min. And similarly we may modify the Definition you gave just now.

Nie. Certainly. 'The distance of a point from a right Line is the length of the perpendicular let fall upon it from the given point.'

Min. What is your next step?

NIEMAND *reads.*

P. 33. *Ded. G.* 'If points be taken along one of the arms of an angle farther and farther from the vertex, their distances from the other arm will at length be greater than any given straight line.'

In proving this we assume as an Axiom that the lesser of two magnitudes of the same kind can be multiplied so as to exceed the greater.

Min. I accept the Axiom and the proof.

NIEMAND *reads.*

P. 34. Ax. 'If one right Line be drawn in the same Plane as another, it cannot first recede from and then approach to the other, neither can it first approach to and then recede from the other on the same side of it.'

Min. Here, then, you assume, as axiomatic, one of the Propositions of Table II. After this, you ought to have no further difficulty in proving Euc. I. 32 and all other properties of Parallels. How do you proceed?

Nie. We prove (p. 34. *Lemma*) that, if two Lines have a common perpendicular, each is equidistant from the other.

Min. What then?

Nie. Next, that any Line intersecting one of these will intersect the other (p. 35).

Min. That, I think, depends on Deduction *G*, at p. 33?

Nie. Yes.

Min. A short, but not very easy, Theorem; and one containing a somewhat intricate diagram. However, it proves the point. What is your next step?

Niemand *reads.*

P. 34. Lemma. 'Through a given point without a given straight Line one and only one straight Line can be drawn in the same Plane with the former, which shall never meet it. Also all the points in each of these straight Lines are equidistant from the other.'

Min. I accept all that.

Nie. We then introduce Euclid's definition of 'Parallels. It is of course now obvious that parallel Lines are equidistant, and that equidistant Lines are parallel.

Min. Certainly.

Nie. We can now, with the help of Euc. I. 27, prove I. 29, and thence I. 32.

Min. No doubt. We see, then, that you propose, as a substitute for Euclid's 12th Axiom, a new Definition, two new Axioms, and what virtually amounts to five new Theorems. In point of 'axiomaticity' I do not think there is much to choose between the two methods. But in point of brevity, clearness, and suitability to a beginner, I give the preference altogether to Euclid's axiom.

The next subject to consider is your practical test, if any, for two given Lines meeting when produced.

Nie. One test is that one of the Lines should meet a Line parallel to the other.

Min. Certainly: and that will suffice in such a case as Euc. I. 44 (Pr. M. p. 60, in this book) though you omit to point out *why* the Lines may be assumed to meet. But what if the diagram does not contain 'a Line parallel to the other'? Look at Pr. (*h*) p. 69, where we are told to make, at the ends of a Line, two angles which are together less than two right angles, and where it is assumed that the Lines, so drawn, will meet. That is, you assume the truth of Euclid's 12th Axiom. And you do the same thing at pp. 70, 123, 143, and 185.

Nie. Euclid's 12th Axiom is easily proved from our Theorems.

Min. No doubt: but you have not done it, and the omission makes a very serious hiatus in your argument. It is not a thing that beginners are at all likely to be able to supply for themselves.

I have no adverse criticisms to make on the general style of the book, which seems clear and well written. Nor is it necessary to discuss the claims of the book to *supersede* Euclid, since the writer makes no such claim, but has been careful (as he states in his preface) to avoid any arrangement incompatible with Euclid's order. The chief novelty in the book is the introduction of the principle of 'equidistance,' which does not seem to me a desirable feature in a book meant for beginners: otherwise it is little else than a modified version of Euclid.

ACT II.

SCENE V.

Treatment of Parallels by revolving Lines.

HENRICI.

'In order that an aggregate of elements may be called a spread, it is necessary that they follow continuously.'—HENRICI'S *Art of Dining*, p. 12.

Nie. I lay before you '*Elementary Geometry: Congruent Figures*,' by OLAUS HENRICI, Ph.D., F.R.S., Professor of Pure Mathematics in University College, London, 1879.

Min. What is your Definition of a Line?

Nie. 'The boundary of a surface or of part of a surface is called a *Line* or a *curve*.' (p. 5.)

Min. Good—'Line,' I presume, meaning 'right Line.' But that throws us back upon 'surface.' Of course *that* is defined correctly?

Nie. I will tell you in a moment. (*He turns over a few pages*) Yes, here it is. 'A surface is the—' (*He gives a perceptible start, stops reading, and turns a few pages back*) Yes, it's all right. 'That which bounds a solid and separates it from other parts of Space is called its *surface*.' (p. 4.)

Min. (*aside*) There is more here than meets the eye! (*Aloud*) You will be good enough to read that *other* Definition of 'surface.'

Nie. (*innocently*) What other Definition?

Min. No evasions, Sir! Read it at once! You know the one I mean.

Nie. (*desperately*) It's only this—'A surface is the path of a moving curve.' (p. 9.) Merely another way of looking at it, you know.

Min. (*contemptuously*) Oh! Merely another way of looking at it, is it? Of course the curve preserves its shape as it moves?

Nie. No doubt.

Min. Now look here. Take this Jargonelle pear—

Nie. Thank you very much. It *is* rather dry work—

Min. Stop! Don't eat it yet! Look at it. Would you call its curvature regular?

Nie. Certainly not: it bulges here and there, in all sorts of queer ways.

Min. Well, now take this bit of wire: bend it into any curve you like, and then move it so that its path may coincide with the surface of the pear.

Nie. (*uneasily*) I cannot do it.

Min. Well, eat it, then. *That* is possible, at all events. So! We start with a Definition which is simply ridiculous! Now for the distinction between 'right Line' and 'curve'—

Nie. Here my client's meaning is not very clear. The first Definition I can find is that of a *curve.* He says (p. 6) 'a point may be moved, and then it will describe a path. This path of a moving point is a curve.'

Min. Surely he does not mean that a point can never move *straight?* He must mean that there are two kinds of curves, 'curved curves,' and 'straight curves'—as the Irish talk of 'tay-tay' and 'coffee-tay.' But, if so, he makes 'Line' and 'curve' synonymous.

Nie. I have looked a little further on, and I find a description of a 'Line,' which seems to limit the word to *bent* Lines. He says (p. 7) 'The notion of a Line may be obtained directly by considering a wire bent into any shape and abstracting all thickness from it.'

Min. So then a 'Line' *must* be bent, though a 'curve' need not be so? Your client has clearly *one* merit—great originality of style!

Nie. Here is another definition of 'curve,' which may be more to your taste. 'A curve is a *one-way spread, with points as elements.*' (p. 10.)

Min. Too much like a dinner *à la Russe.* I don't like 'spread' at all.

Nie. He illustrates his use of 'spread' by applying it to other subjects. For instance, 'a musical tone allows of variations which form a two-way spread, with different degrees of intensity and of pitch as elements.' (p. 12.)

Min. That explains the phrase 'too-tooing on a flute.' How simple and intelligible all this must be to boys just beginning Geometry! But I am still waiting for a definition of 'right Line.'

Nie. (*after turning over several pages*) I have found it at last—after passing over a good deal about 'continuity' and 'space' and 'congruence.' We say (p. 17) 'If we suspend a weight by a string, the string becomes stretched; and we say it is straight.'

Min. That will serve very well to give a *notion* of 'straight.' For a *working* definition we require of course some practical test, such as 'two straight Lines cannot enclose a space.'

Nie. We have that. At p. 20 we give you 'Axiom IV. Through two points always one, and only one, Line can be drawn.' And at p. 18 we at last distinguish 'Line' and 'curve.' 'A straight Line will in future be called a *Line* simply. All other Lines will be called *curved Lines*, or *curves*.'

Min. Better late than never: though it makes wild work of your former theory—in which you got the notion of 'Line' from a bent wire, and of 'curve' from the path of a moving point. Now for the Definition of 'angle.'

Nie. (*after turning the leaves backwards and forwards for some time, begins to read in an unsteady voice*) 'The part of a pencil of half-rays, described by a half-ray on turning about its end point from one position to another, is called an angle.' (p. 47.)

Min. So you reject the notion of 'inclination' (or rather 'declination')? Well! This *is* an innovation! We must investigate it thoroughly. You mean by 'half-ray,' I presume, what Euclid calls 'a Line terminated in one direction but not in the other'?

Nie. Certainly.

Min. Now what is a 'pencil'?

Nie. 'The aggregate of all Lines in a plane which pass through a given point.' (p. 38.)

Min. Aha! And where will you get your angular

magnitude, I should like to know? What kinds of magnitude is a Line capable of possessing?

Nie. Length only, of course.

Min. Two Lines?

Nie. (*uneasily*) Length only.

Min. A million?

Nie. (*more uneasily*) Length only.

Min. A pencil?

Nie. (*faintly*) Spare me!

Min. So much for the *quality* of your angular magnitude! Now for its *quantity*. What is the length of one of these half-rays?

Nie. Infinite, of course.

Min. And the aggregate length of all the half-rays in your 'angle' cannot well be less. Thus we may deduce a truly delightful definition of angular magnitude. 'As to *quality*, it is linear. As to *quantity*, it is infinite'!

Nie. (*writhes, but says nothing*).

Min. Will you not throw up your brief?

Nie. Not yet: I must fight it out.

Min. Then we must review this marvellous book 'to the bitter end.' What have you to say about 'right angles'?

Nie. We have 'angles of rotation' and 'angles of continuation' (p. 48); and the axiom 'all angles of rotation are equal' (p. 49) as a substitute for 'all right angles are equal.'

Min. It is a practicable method, but not so suitable for beginners as Euclid's. This matter I have already discussed (see p. 74). And now for the subject of Parallels.

Nie. We have Playfair's Axiom (or rather its equivalent) 'Through a given point only one Line can be drawn parallel to a given Line' (p. 68), but this we do not simply lay down *as* an Axiom. We lead up to it by two or three pages of reasoning.

Min. This is *most* interesting! Let us examine the argument minutely. A logical *proof* of that Axiom would be perhaps the greatest advance ever made in the subject since the days of Euclid.

Nie. 'Two indefinite Lines in a Plane may intersect, as we have seen. We shall now consider the possibility of there being such Lines which do not intersect.' (p. 65.)

Min. *That*, of course, you can easily prove, without appealing to any disputable Axiom. It is simply Euc. I. 27. Do you prove it in Euclid's way?

Nie. Not exactly. Our argument is quite different from Euclid's: and we come to *two* conclusions—one being the real existence of Parallels, the other the equivalent of Playfair's Axiom.

Min. I very much doubt your proving the first by any simpler method than Euclid's: and as to proving the second, by any method *at all*, without assuming some disputable Axiom, I defy you to do it! However, let us hear your argument.

Nie. We take a Line, and a point without it: and from the point we draw two 'half-rays' intersecting the line. These half-rays we then turn about the point, in opposite directions, until they cease to intersect the Line.

And then we proceed to consider where their 'productions' have got to.

Min. Like 'little Bo-peep,' you are anxious about their '*tails*' in fact; taking their 'heads' to be the ends which at first intersected the given Line.

Nie. We say that there are only three conceivable cases: one, where the *tails* fall next to the given Line; another, where the *heads* fall next to it; the third, where the tail of each coincides with the head of the other.

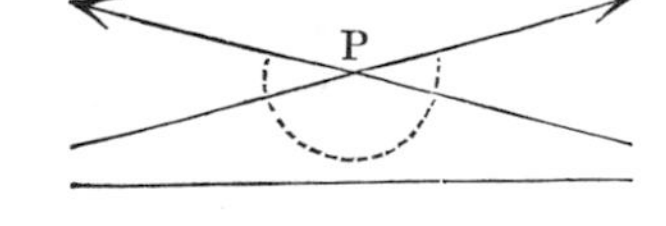

Min. I admit all that.

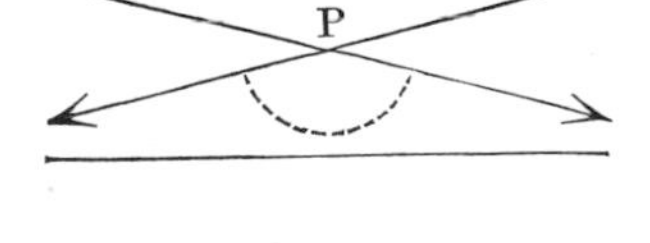

Nie. The first case we say is inadmissible because, if it were true, any Line through P, lying within the angle formed by the head of one ray and the tail of the other, would cut the given Line both ways.

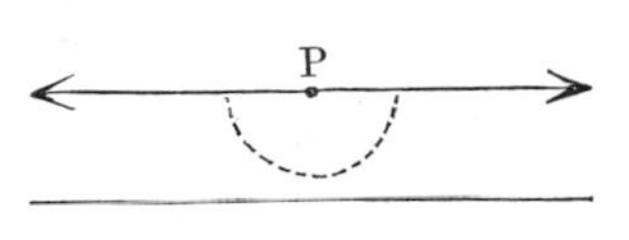

Min. A *reductio ad absurdum*, no doubt; but it only holds good on the supposition that you *can* draw Lines through P, so as to lie within that angle. But this supposition requires a finite angle. If we suppose that, the moment one ray begins to revolve so as to bring its head nearer to the given line, it instantly coincides with the other ray, head with tail and tail with head, it will not then be *possible* to draw any such Line as you suggest: and *then* where is your *reductio ad absurdum*?

Nie. We do not seem to have noticed that case.

Min. In point of fact, your *three* cases are really *five.*

Before going any further let us have them all clearly stated.

We assume, in all three figures, that the ray-heads, as drawn, do *not* intersect the given line; but that either of them would, if it began to revolve towards the given Line, instantly intersect it. In other words we assume that any half-ray, drawn from P in the dotted angular space, *would* intersect the given Line: but that any half-ray, drawn from P in the undotted angular space, as well as the two ray-heads which limit that angular space, would *not* intersect it. And as to the ray-tails, it is obvious that in fig. 1 they *do* intersect the given Line, but in figs. 2, 3, they do *not* do so.

Nie. That is all clear enough.

Min. Then these are the five cases:—

(α) Figure 1. The head of each ray must revolve downwards through a *finite* angle before it can coincide with the tail of the other ray.

(β) Same figure. The head of each ray, on beginning to revolve downwards, *instantly* coincides with the tail of the other ray.

(γ) Figure 2. The head of each ray must revolve upwards through a *finite* angle before it can coincide with the tail of the other ray.

(δ) Same figure. The head of each ray, on beginning to revolve upwards, *instantly* coincides with the tail of the other ray.

(ϵ) Figure 3.

These five cases suggest a few observations.

In case (α) a number of Lines may be drawn through

P, in the angular space contained between the head of one ray and the tail of the other: and all such Lines will intersect the given line *both* ways.

In case (β) this absurdity does not arise: all Lines, through *P*, intersect the given Line one way or the other: there is no instance of a Line intersecting it *both* ways, nor of one wholly separate from it.

In case (γ) a number of Lines may be drawn as in case (α): and all such Lines will be wholly separate from the given Line.

In case (δ) the two rays themselves, as drawn in the figure, are wholly separate from the given Line: but no other such Line can be drawn through *P*.

In case (ϵ) there is only *one* Line through *P* wholly separate from the given Line.

Now let us hear what you make of these five cases.

Nie. We exclude case (α), as I told you just now, by a *reductio ad absurdum.* Case (β) we have failed to notice.

Min. True: but it *can* be excluded by Euc. I. 27: so that if you can manage, by pure reasoning, from ordinary Axioms, and without assuming any disputable Axiom, to exclude cases (γ) and (δ), you will have achieved what geometricians have been vainly trying to do for the last two thousand years!

Nie. We go on thus. 'But our Axioms are not sufficient to decide which of the remaining two cases actually does occur.' (p. 67.)

Min. Or rather 'the remaining *three* cases.'

Nie. 'In looking at the figures the reader will at once

feel that the third case' (we mean your 'case (ϵ)') 'is the true one.'

Min. An appeal to sentiment! What if the reader *doesn't* feel it?

Nie. 'But this cannot be considered decisive;'

Min. It cannot.

Nie. 'for the two Lines may include a very small angle —'

Min. Aye, or even a large one.

Nie. 'that is, they may very nearly coincide without actually doing so. Or it may be that sometimes the one, sometimes the other, happens, according as we take the point P at a smaller or greater distance from the Line.'

Min. That seems a fair statement of the difficulty. And now, how are you going to grapple with it?

Nie. 'The only way of settling this point is to make an assumption, and to see whether the consequences drawn from it do or do not agree with our experience.'

Min. If you find a consequence *not* agreeing with experience, you may of course conclude that your assumption was false; but, if it *does* agree, what then?

Nie. Nothing, I fear, unless you can prove that this is the case with *one* assumption only, and that all other possible assumptions lead to absurd results.

Min. Exactly so. If, then, you want to prove case (ϵ), your logical course is to assume case (γ) as true, and from that assumption to deduce some consequence which is evidently contrary to experience. And then to exclude case (δ) by a similar argument. Is that your method?

Nie. Well, hardly. We say 'The assumption to be

made is, that the third case' (i.e. case (ϵ)) only happens, and this will give us a new axiom.' (p. 67.)

Min. You may assume it as an *axiom*, if you like. Then you will merely be in the same boat with Playfair. But if you are going to discuss the *consequences* of its being true, and get anything out of *that*, look to your feet! There are pitfalls about!

Nie. 'In the second case' (i.e. case (γ)) 'we should have to—'

Min. Oho! Then it is case (γ), after all, that you are provisionally assuming as true?

Nie. Apparently so.

Min. Well, go on. You are on the right track now.

Nie. In this case we should have to turn the ray 'through a finite angle' before its tail would cut the given Line: 'or there would be an indefinite number of Lines through P which do not cut' it. (p. 68.)

Min. What do you mean by 'or'? That one result would follow, *or* the other, but not both?

Nie. We mean that the two results are equivalent.

Min. Then you should say 'that is.' 'Or' is misleading. However, I grant you that this consequence *would* follow, if case (γ) were true. What then? Is there any obvious absurdity in such a consequence?

Nie. *That* we do not assert. We merely make the remark—and we now proceed to case (ϵ).

Min. A weak and pointless remark: but let that pass. Do you omit case (δ)?

Nie. We do. We proceed thus. 'But in the third case (i.e. in case (ϵ)) there would be only *one* Line through

P which does not cut' the given Line. 'As soon as we turn this Line about *P* it would meet it to the right or to the left.'

Min. Certainly. And what then? Do you expect me to admit that, because case (ϵ) would lead to a consequence not obviously absurd, therefore it is *the* case which always happens, to the exclusion of cases (γ) and (δ)?

Nie. (*hesitatingly*) Well, I think that *is* what we expect. But we first deduce the real existence of Parallels. 'Thus we are led to the conclusion that there exist Lines in a Plane which, though both be unlimited, do not meet. Such Lines are called *parallel*.'

Min. Oh most lame and impotent conclusion! After all these magnificent Catherine-wheels of revolving half-rays, to deduce Euc. I. 27! And even *this* wretched result you have no right to. Just consider what your argument has been. There are five conceivable cases, (α), (β), (γ), (δ), and (ϵ). If (α) or (β) were true, *no* Line could be drawn, through *P*, parallel to the given Line: if (γ), *many* such Lines could be drawn: if (δ), *two* such Lines: if (ϵ), *one* such Line. Now what have you proved? Positively nothing whatever but this—that case (α) would lead to an absurd result. You leave me perfectly free to range about among the other four cases, one of which, (β), denies the real existence of Parallels, which existence you tell me you have *proved!* And so, for the 'long course of logical reasoning' which you object to so much in Euclid, you substitute a *short* course of *il*logical reasoning! But you deduce another conclusion, do you not?

Nie. Yes, one other. 'The assumption mentioned in § 113' (the assumption that case (ε) is the only true one) 'may now be stated thus:—Axiom VI. *Through a given point only one Line can be drawn parallel to a given Line.*'

Min. May it indeed? And why 'now' rather than three pages back? Is there a single word, in all this argument, which tends to show that case (ε) is—I will not say certainly true, but—even fairly probable?

Nie. (*cautiously*) I will not assert that there is.

Min. In point of fact the odds are exactly three to one against it—since you have only excluded *one* of the five cases, and the other four are, for anything we know to the contrary, equally probable.

Nie. I will not dispute it.

Min. Well! Then it only remains to say that your attempted *proof* of Playfair's Axiom is an utter failure. Anything more hopelessly illogical I have *never* met with, not even in Cooley—and that is saying a great deal!

Nie. I confess I do not see my way to defending this proof. But even if we abandon the whole of it, we are no worse off than any other writer who assumes Playfair's Axiom.

Min. That I quite admit.

Nie. And then, my client instructs me to plead, this Manual (*handing it to Minos*) being so distinctly better than Euclid's in every other particular—

Min. Gently, gently! You are anticipating. I have not yet had my general survey of the book.

Nie. (*refilling his pipe*) Well, let us have it then.

Min. I will begin with the general remark that the

first 151 pages of this book (the rest of it going beyond the limits of Euc. I, II) contain (excluding 7 pages on Logic and 22 pages of Exercises) 122 pages of text, which I presume the learner is expected to master.

Nie. A great deal of that is merely explanatory.

Min. True: but even omitting all that, we have, of Definitions, 80: and of Theorems, 145. And when the unfortunate learner has mastered all these—more than there are in Euclid's first six Books—he finds he has learned no more Euclid than Props. 1 to 34!

Nie. But he will have learned a good deal that is *not* in Euclid.

Min. Undoubtedly: and it would have been easy to crowd in twice as many Theorems as Mr. Henrici has done, without passing Prop. 34. I believe the subject to be practically inexhaustible. But fancy having to master 145 Theorems before even hearing of so important a one as Prop. 47!

Nie. If all the new matter is *good,* it is a poor objection to raise that there is too much of it.

Min. You think the *quantity* unassailable? Well, let us test its *quality* a little, then.

The book begins with a page or two of very general considerations. Time and Force, Kinetics and Kinematics, Chemistry and Biology, cross the stage in a grand but shadowy procession. Then when the pupil has been sufficiently crushed by the spectacle of how much there is to know, we allow him, little by little, to contract his view: till at last we condescend to contemplate so trifling an entity as Infinite Space.

And here I notice a singular mental process. 'Two material bodies,' we are told, 'cannot occupy the same space. We are thus led to recognise a third property common to all bodies: every body has *position*.' (p. 3.) The word '*thus*' is what I want to call your special attention to: for I confess *I* can see no such sequence of thought as it would seem to imply. Suppose bodies *could* occupy the same space: wouldn't they have 'position' just as much as if they couldn't? Does an orange—to take the favourite logical entity—lose its position because another orange most uncivilly insists on permeating it and occupying the same portion of Space? But if not, what is the meaning of 'thus'? As Artemus Ward would say, 'why this thusness?'

Nie. I can't say.

Min. A little further on I find a 'therefore' which is equally shadowy. The writer's logical ideas—in spite of his actually introducing a 'Digression on Logic'—are, I fear, a little vague. He says 'If we bring different points together into the same position, they will never give us anything but a point; we never obtain any extension. We cannot, therefore, say that Space is made up of points' (p. 6). I venture to say that there is no such sequence as 'therefore' seems to imply: he has made the whole argument null and void by using the words 'into the same position.'

Nie. I do not understand you.

Min. I will put it in another way. The *real* reason why you cannot construct Space of points is that they have no *size*: if they *had* size you *could* do it. But, under the condition here laid down—of bringing them 'together into the same position'—you make the thing impossible, whether they have size or not.

I have often found it the best way for exhibiting the unsoundness of an argument, to make another exactly like it, but leading to an absurd conclusion. I will try it here. You grant that a cubic foot *can* be made up of cubic inches?

Nic. Certainly.

Min. Well, I will prove to you that it *cannot*; and I will do so by an argument just as good as Mr. Henrici's. 'If we bring different cubic inches together into the same position they will never give us anything but a cubic inch; we never obtain any extension—'

Nic. That won't do! You have the 'extension' of one cubic inch.

Min. Yes, but you had that to begin with. You don't 'obtain' any extension by squeezing in *other* cubic inches, do you?

Nic. No, I suppose not.

Min. Then the argument is sound so far. And now comes my triumphant conclusion, *à la Henrici.* 'We cannot, *therefore*, say that a cubic foot is made up of cubic inches.'

Nic. I see your meaning now. I give up the words 'into the same position.'

Min. I haven't quite done with points yet. I find an assertion that they never jump. Do you think that arises from their having 'position,' which they feel might be compromised by such conduct?

Nie. I cannot tell without hearing the passage read.

Min. It is this:—'A point, in changing its position on a curve, passes, in moving from one position to another, through all intermediate positions. It does not move by jumps.' (p. 12.)

Nie. That is quite true.

Min. Tell me, then—is every centre of gravity a point?

Nie. Certainly.

Min. Let us now consider the centre of gravity of a flea. Does it—

Nie. (*indignantly*) Another word, and I shall vanish! I cannot waste a night on such trivialities.

Min. Forgive me. I drop the flea. My next remark shall be serious. I wish to point out to you the illogical *tone* of the book. I do not say that the instances I am going to give are crucial or fatal to the argument. But, however unimportant, and however easily corrected, they will, I think, justify me in asking 'Is a text-book, which contains such loosely reasoned arguments as these, to be trusted?'

My first selection is § 52, p. 23. For brevity's sake I shall omit superfluous words. The passages in parentheses are interpolations of my own.

(*see Henrici, p.* 23.)

'If we conceive a Plane (and a point *A* chosen anywhere in Space; then, either the Plane already passes through *A*, or) we may move it until a point on it comes to *A*, which has been chosen anywhere in Space. (If we now fix a second point *B*; then, either the Plane already passes through *B*, or) if we keep *A* fixed we may turn the Plane about it, until the Plane comes to pass also through *B*, likewise chosen arbitrarily in Space. (If we now fix a third point *C*; then, either the Plane already passes through *C*, or) we may still move the Plane, as only two points of it are fixed, by turning it about the Line joining them, until the Plane passes through *C*, chosen arbitrarily, like *A* and *B*. Thus it appears that we may place a Plane so as to pass through three points, *A*, *B*, *C*, chosen anywhere in Space.'

You accept that, interpolations and all?

Nie. Certainly.

Min. Omit the interpolations, and what do you say of it then?

Nie. It remains true. The three successive movings do no harm, but they are not always *necessary*.

Min. Would this statement be correct? 'Three "movings" are *generally* necessary: but there are three exceptions. If the Plane at first passes through *A*, the *first* "moving" is unnecessary; if, after being made to pass through *A*, it be found to pass through *B* also, the *second* "moving" is unnecessary; and if, after being made to pass through *A* and *B*, it be found to pass through *C* also, the *third* "moving" is unnecessary'?

Nie. Certainly.

Min. You would not, on finding some one 'moving' unnecessary, call it 'an open question' whether the result were attainable?

Nie. What? When it is already attained? By no means.

Min. Now read this, at p. 23.

(*hands the book*)

'But if *C* happens to lie on the Line joining *A* and *B*, then a Plane through *A* and *B*, which did not pass through *C*, could never be made to pass through *C* by being rotated about *A* and *B*; for if it did contain *C* in one position, it would contain it in all positions, as this point would remain fixed during rotation.' What do you say to that?

Nie. Well, it is his way of discussing your third exception. Of course, when he talks of 'a Plane through *A* and *B*, which did not pass through *C*,' he is describing a nonentity: but it is all logical as an argument.

Min. What kind of argument?

Nie. (*doubtfully*) I should call it a—kind of—*Reductio ad Absurdum*.

Min. I don't wonder at your hesitation. A thoughtful boy might read it thus:—'then a Plane through *A* and *B*, which did not pass through *C* (but no such Plane can exist!), could never be made to pass through *C* by being rotated about *A* and *B* (why, it needs no 'making'!); for if it did contain *C* in one position (which it does!), it would contain it in all positions (which also it does!)'

You and I can recognise the *Reductio ad Absurdum*—though so abnormal and hideous—which the writer intends. But what do you think would be the effect, on a thoughtful boy, of a course of such arguments, where he is expected to accept as *data* what he knows to be absurd, and to recognise as an absurdity what he knows to be a necessary truth?

Nie. At first, Mania: ultimately, Dementia.

Min. Now read Mr. Henrici's deduction from this fearful argument, at p. 24.

'We ought, therefore, to limit the conclusion arrived at as follows:—Through three points *which do not lie in a Line* we may always pass a Plane. Whether a Plane may be drawn through three points which do lie in a Line, remains for the moment an open question.'

Are you prepared to back that statement? *Is* it an 'open question'?

Nie. I cannot say that it is.

Min. Now here is a most curious bit of bad Logic. (*reads*) 'If two Planes have two points, *A* and *B*, in common, they must necessarily have more points in common. For, since each extends continuously without limit, a point moving in the one Plane through *A* or *B* will cross the other Plane at this point;' (p. 25.)

I pause to ask—will it *necessarily* do so? How if it moved along their Line of intersection?

Nie. That is an exception, I grant.

Min. (*reads*) 'hence one Plane will lie partly on the one and partly on the other side of the second Plane. They must therefore intersect.'

Now the conclusion—that the Planes intersect—is undoubtedly true, so long as we assume that, by 'two planes,' the writer means 'two *different* Planes.' But does it follow from the *premisses*? Have the words 'hence' and 'therefore' any logical value?

Nie. I fear not.

Min. At p. 74 I observe 'If two Lines be each perpendicular to a third, they will be parallel to one another.' This is not true. They might be coincidental. The same mistake is made in p. 75.

Now comes a wonderful specimen of slipshod writing. 'We understand by the angles of a Polygon those angles of which the part near the vertex lies within the Polygon.' Does not this oblige us to contemplate an angle as consisting of *two parts*—one 'near the vertex,' the other further off?

Nie. Undoubtedly.

Min. And if either part were gone, the angle would be less?

Nie. (*uneasily*) It would seem so.

Min. And this might be effected by shortening the Lines, so that they would not reach beyond the region 'near the vertex'?

Nie. I fear you have got us into a corner. Be merciful!

Min. You mean that I have driven you into 'that part of an angle which lies near the vertex.' Well, you may come out now. We will seek 'fresh fields and pastures new.'

At pages 91 to 96 I find no less than forty-six theorems on Symmetry, arranged in two columns—one headed 'Axial Symmetry,' the other 'Central Symmetry.' Here is a specimen pair, at p. 95.

'Corresponding Polygons are congruent but of opposite sense.'	'Corresponding Polygons are congruent and of like sense.'

I hardly know which to pity most—the master who has to teach these Theorems, or the boy who has to learn them!

But I have neither the 'one-way spread with moments as elements' nor the 'three-way spread with points as elements' to—

Nie. (*gasping*) What *are* you talking about?

Min. Excuse me. I fear I am getting demoralised. I meant to say—I have neither the time nor the space to criticise this book throughout.

I will, however, try to sum up its faults in a general description.

'Olla Podrida' is perhaps the best name for it, its contents are so hopelessly jumbled together. Most of the Axioms, and all the Theorems, are without numbers, and, as there is no index, the difficulty of finding them when wanted is obvious: and none the less that they are imbedded in oceans of 'padding.' Dip into the book anywhere, and you find yourself in the midst of some discursive talk, which perhaps culminates in an Axiom. Then perhaps comes a Definition. Then comes a little more talk, which, after appealing to sentiment, or probability, or some other motive degrading to Pure Mathematics, gradually becomes more and more logical, and at last warms into a regular proof—but of what? The reader has no warning as to *what* is to be proved. Unsuspectingly he glides on with the stream, till with a crash he is landed on an enunciation, and finds himself committed to an entire Theorem. This singular writer always reserves the enunciation for the *end* of the Proposition. It may be prejudice, but I cannot help thinking that Euclid's plan—of first clearly stating what he is going to prove and then proving it—is to be preferred to this conjurer's trick of 'forcing a card.'

The book is, I think, *very* hard for beginners to master: the majority of the new Theorems are much more fitted for 'exercises,' than to be embodied in a text-book: and, to crown all, the ambitious attempt to construct a *proof* of Playfair's Axiom is, as we have seen, a lamentable failure.

I think I cannot better conclude my review of this book than by giving you, in two parallel columns, Euclid's Props. I. 18, 19, and Mr. Henrici's proposed substitute for them, at p. 107.

(*turn over.*)

EUCLID.

The greater side of a Triangle is opposite to the greater angle.

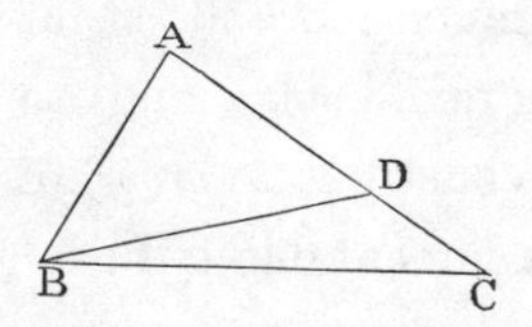

Let ABC be a triangle having $AC > AB$: then shall the angle ABC be > the angle C.

From AC cut off AD equal to AB; and draw BD.

Then, ∵ $AB = AD$, ∴ the angle $ABD =$ the angle ADB;

but the angle ADB is exterior to the Triangle BCD, and ∴ < the angle C;

∴ the angle ABD also > the angle C;

much more is the angle $ABC >$ the angle C. Q. E. D.

HENRICI.

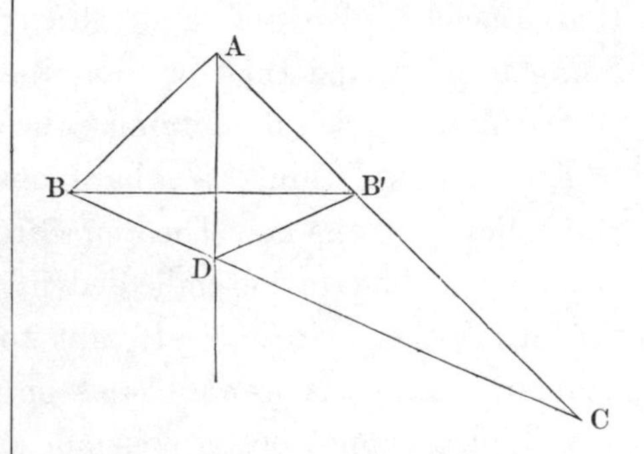

Let us now suppose a Triangle ABC, in which the bisector of the angle BAC is not an axis of symmetry. Then the contra-positive form of the theorem of § 162 tells us that AB is not equal to AC, that the angle B is not equal to the angle C, and that the bisector AD of the angle A is not perpendicular to BC, and hence, that the two angles ADB and ADC are unequal. Between these angles there exists the relation

'the angle $ABC +$ the angle BDA = the angle $C +$ the angle CDA,'

for each sum makes with half the angle A an angle of continuation. Hence it follows that, if the angle $ABC >$ the angle C, the angle $BDA <$ the angle CDA.

The greater angle of a Triangle is opposite to the greater side.

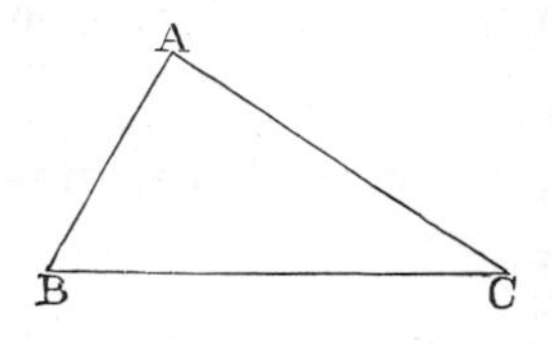

Let ABC be a Triangle having the $\angle B >$ the angle C: then shall AC be $> AB$.

For if not, it must be equal or less.

It is not equal, for then the angle B would $=$ the angle C.

It is not less, for then the angle B would $<$ the angle C.

$\therefore AC > AB$.

Q. E. D.

If we now fold the figure along AD, then AB will fall along AC; and B will fall between A and C if we suppose that AB is the shorter of the two unequal Lines AB and AC. The line DB therefore takes the position DB' within the angle ADC. But the angle $AB'D$, which $=$ the angle ABC, is exterior to the triangle DCB', and $\therefore >$ the angle C.

Conversely, if the angle $ADB <$ the angle ADC, the line DB will fall within the angle ADC, and $\therefore B$ will fall between A and C; that is, AB will be less than AC. This always happens (see above) if the angle $ABC >$ the angle C, for then the angle $BDA <$ the angle ADC.

Theorem. *In every Triangle the greater side is opposite to the greater angle*, and conversely, *the greater angle is opposite to the greater side.*

Now, if you could get some schoolmaster—one who had no bias whatever in favour of Euclid or of Henrici—to teach these two columns (one containing 169, the other 282 words) to two ordinary boys of equal intelligence, or rather of equal stupidity, what result would you expect?

Nie. (*with a cunning smile*) I don't think I could find such a schoolmaster.

Min. Ah, crafty man! You evade the question! I can't resist giving you just *one* more tit-bit—the definition of a Square, at p. 123.

'A quadrilateral which is a kite, a symmetrical trapezium, and a parallelogram is a Square'!

And now, farewell, Henrici! 'Euclid, with all thy faults, I love thee still!' Indeed I might say 'with twice thy faults,' or 'with thrice thy faults,' if the alternative be Henrici! (*returns the book, which Niemand receives in solemn silence.*)

ACT II.

SCENE VI.

Treatment of Parallels by direction.

---

§ 1. WILSON.

'There is moreover a logic besides that of mere reasoning.'
WILSON, *Pref.* p. xiii.

Nie. You have made but short work of four of the five methods of treating Parallels.

Min. We shall have all the more time to give to the somewhat intricate subject of Direction.

Nie. I lay on the table '*Elementary Geometry,*' by J. M. WILSON, M.A., late Fellow of St. John's College, Cambridge, late Mathematical Master of Rugby School, now Head Master of Clifton College. The second edition, 1869. And I warn you to be careful how you criticise it, as it is already adopted in several schools.

Min. Tant pis pour les écoles. So you and your client deliberately propose to supersede Euclid as a text-book?

Nie. 'I am of opinion that the time is come for making an effort to supplant Euclid in our schools and universities.' (Pref. p. xiv.)

Min. It will be necessary, considering how great a change you are advocating, to examine your book *very* minutely and critically.

Nie. With all my heart. I hope you will show, in your review, 'the spirit without the prejudices of a geometrician.' (Pref. p. xv.)

Min. We will begin with the Right Line. And first, let me ask, how do you define it?

Nie. As 'a Line which has the same direction at all parts of its length.' (p. 3.)

Min. You do not, I think, make any practical use of that as a test, any more than Euclid does of the property of lying evenly as to points on it?

Nie. No, we do not.

Min. You construct and test it as in Euclid, I believe? And you have his Axiom that 'two straight Lines cannot enclose a space?'

Nie. Yes, but we extend it. Euclid asserts, in effect, that two Lines, which coincide in two points, coincide *between* those points: *we* say they 'coincide wholly,' which includes coincidence *beyond* those points.

Min. Euclid tacitly assumes that.

Nie. Yes, but he has not expressed it.

Min. I think the addition a good one. Have you any other Axioms about it?

Nie. Yes, 'that a straight Line marks the shortest distance between any two of its points.' (p. 5. Ax. 1.)

Min. That I have already fully discussed in reviewing M. Legendre's book (see p. 55).

Nie. We have also 'A Line may be conceived as transferred from any position to any other position, its magnitude being unaltered.' (p. 5. Ax. 3.)

Min. True of *any* geometrical magnitude: but hardly worth stating, I think. I have now to ask you how you define an Angle?

Nie. 'Two straight Lines that meet one another form an angle at the point where they meet.' (p. 5.)

Min. Do you mean that they form it '*at* the point' and nowhere else?

Nie. I suppose so.

Min. I fear you allow your angle no magnitude, if you limit its existence to so small a locality!

Nie. Well, we *don't* mean 'nowhere else.'

Min. (*meditatively*) You mean '*at* the point—and *somewhere* else.' *Where* else, if you please?

Nie. We mean—we don't quite know why we put in the words at all. Let us say 'Two straight Lines that meet one another form an angle.'

Min. Very well. It hardly tells us what an angle *is*, and, so far, it is inferior to Euclid's Definition: but it may pass. Do you put any limit to the *size* of an angle?

Nie. We have not named any, but the largest here treated of is what we call 'one revolution.'

Min. You admit reëntrant angles then?

Nie. Yes.

Min. Then your Definition only states half the truth: you should have said 'form *two* angles.'

Nie. That would be true, no doubt.

Min. But this extension of limit will require several modifications in Euclid's language: for instance, what is your Definition of an obtuse angle?

NIEMAND *reads.*

P. 8. Def. 13. 'An *obtuse angle* is one which is greater than a right angle.'

Min. So you tumble headlong into the very first pitfall you come across! Why, that includes such angles as 180° and 360°. You would teach your pupil, I suppose, that one portion of a straight Line makes an obtuse angle with the other, and that every straight Line has an obtuse angle at each end of it!

Nie. It is an oversight—of course we ought to have added 'but less than two right angles.'

Min. A very palpable oversight. I fear we shall find more as we go on. What Axioms have you about angles?

NIEMAND *reads.*

P. 5. Ax. 4. 'An angle may be conceived as transferred to any other position, its magnitude being unaltered.'

Min. Hardly worth stating. Proceed.

NIEMAND *reads.*

P. 5. Ax. 5. 'Angles are equal when they could be placed on one another so that their vertices would coincide in position, and their arms in direction.'

Min. 'Placed on one another'! Did you ever see the child's game, where a pile of four hands is made on the table, and each player tries to have a hand at the top of the pile?

Nie. I know the game.

Min. Well, did you ever see both players succeed at once?

Nie. No.

Min. Whenever that feat is achieved, you may *then* expect to be able to place two angles 'on one another'! You have hardly, I think, grasped the physical fact that, when one of two things is *on* the other, the second is *underneath* the first. But perhaps I am hypercritical. Let us try an example of your Axiom: let us place an angle of 90° on one of 270°. I think I could get the vertices and arms to coincide in the way you describe.

Nie. But the one angle would not be *on* the other; one would extend round one-fourth of the circle, and the other round the remaining three-fourths.

Min. Then, after all, the angle is a mysterious entity, which extends from one of the Lines to the other? That is much the same as Euclid's Definition. Let us now take your definition of a Right Angle.

Nie. We first define 'one revolution,' which is the angle described by a Line revolving, about one extremity, round into its original position.

Min. That is clear enough.

Nie. We then say (p. 7. Def. 9) 'When it coincides with what was initially its continuation, it has described *half a revolution*, and the angle it has then described is called *a straight angle*.'

Min. How do you know that it has described *half a revolution*?

Nie. Well, it is not difficult to prove. Let that portion of the Plane, through which it has revolved, be rolled over, using as an axis the arm (in its initial position) and its continuation, until it falls upon the other portion of the Plane. The two angular magnitudes will now together make up 'one revolution': therefore each is 'half a revolution.'

Min. A proof, I grant: but you are *very* sanguine if you expect beginners in the subject to supply it for themselves.

Nie. It is an omission, we admit.

Min. And then 'a straight angle'! 'Straight' is necessarily unbending: while 'angle' is from ἄγκος, 'a bend or hook': so that your phrase is exactly equivalent to 'an unbending bend'! In 'the Bairnslea Foaks' Almanack' I once read of 'a mad chap' who spent six weeks 'a-trying to maäk a straät hook': but he failed. He ought to have studied your book. Have you Euclid's Axiom 'all right angles are equal'?

Nie. We deduce it from 'all straight angles are equal': and that we prove by applying one straight angle to another.

Min. That is all very well, though I cannot think 'straight angles' a valuable contribution to the subject. I will now ask you to state your method of treating Pairs of Lines, as far as your proof of Euc. I. 32.

Nie. To do that we shall of course require parallel Lines: and, as our definition of them is 'Lines having the same direction,' we must begin by discussing direction.

Min. Undoubtedly. How do you define direction?

Nie. Well, we have not attempted *that.* The idea seemed to us to be too elementary for definition. But let me read you what we have said about it.

Reads.

P. 2. Def. 2. 'A *geometrical Line* has position, and length, and at every point of it it has direction . . . .'

P. 3. Def. 4. 'A *straight Line* is a Line which has the *same* direction at all parts of its length. It has also the opposite direction . . . . A straight Line may be conceived as generated by a point moving always in the *same* direction.'

I will next quote what we have said about two Lines having 'the same direction' and 'different directions.'

Min. We will take that presently: I have a good deal to say first as to what you have read. I gather that you consider direction to be a *property* of a geometrical entity, but not itself an entity?

Nie. Just so.

Min. And you ascribe this property to a Line, and also to the motion of a point?

Nie. We do.

Min. For simplicity's sake, we will omit all notice of curved Lines, etc., and will confine ourselves to straight Lines and rectilinear motion, so that in future, when I use the word 'Line,' I shall mean 'straight Line.' Now may we not give a notion of 'direction' by saying—that a moving point must move in a certain 'direction'—that, if two points, starting from a state of coincidence, move along two equal straight Lines which do not coincide (so that their movements are alike in point of departure, and in magnitude), that quality of each movement, which makes it differ from the other, is its 'direction'—and similarly that, if two equal straight Lines are terminated at the same point, but do not coincide, that quality of each which makes it differ from the other, is its 'direction' from the common point?

Nie. It is all very true: but you are using 'straight Line' to help you in defining 'direction.' *We*, on the contrary, consider 'direction' as the more elementary idea of the two, and use it in defining 'straight Line.' But we clearly agree as to the meanings of both expressions.

Min. I am satisfied with that admission. Now as to the phrase 'the same direction,' which you have used in reference to a single Line and the motion of a single point. May we not say that portions of the same Line have 'the same direction' as one another? And that, if a point moves along a Line without turning back, its motion at one instant is in 'the same direction' as its motion at another instant?

Nie. Yes. That expresses our meaning in other language.

Min. I have altered the language in order to bring out clearly the fact that, in using the phrase 'the same direction,' we are really contemplating *two* Lines, or *two* motions. We have now got (considering 'straight Line' as an understood phrase) accurate geometrical Definitions of at least *two* uses of the phrase. And to these we may add a third, viz. that two coincident Lines have 'the same direction.'

Nie. Certainly, for they are one and the same Line.

Min. And you intend, I suppose, to use the word 'different' as equivalent to 'not-same.'

Nie. Yes.

Min. So that if we have, for instance, two equal Lines terminated at the same point, but not coinciding, we say that they have 'different directions'?

Nie. Yes, with one exception. If they are portions of one and the same infinite Line, we say that they have 'opposite directions.' Remember that we said, of a Line, 'it has also the opposite direction.'

Min. You did so: but, since 'same' and 'different' are contradictory epithets, they must together comprise the whole genus of 'pairs of directions.' Under which heading will you put 'opposite directions'?

Nie. No doubt, strictly speaking, 'opposite directions' are a particular kind of 'different directions.' But we shall have endless confusion if we include them in that class. We wish to avoid the use of the word 'opposite' altogether, and to mean, by 'different directions,' all kinds of directions that are not the same, with the exception of 'opposite.'

Min. It is a most desirable arrangement: but you have not clearly stated it in your book. Tell me whether you agree in this statement of the matter. Every Line has a pair of directions, opposite to each other. And if two Lines be said to have 'the same direction,' we must understand 'the same *pair* of directions'; and if they be said to have 'different directions,' we must understand 'different *pairs* of directions.' And even this is not enough: for suppose I draw, on the map of England, a straight Line joining London and York; I may say 'This Line has a pair of directions, the first being "London-to-York" and the second "York-to-London."' I will now place another Line upon this, and *its* pair of directions shall be, first "York-to-London" and second "London-to-York." Then it has a different first-direction from the former Line, and also a different second-direction: that is, it has a 'different pair of directions.' Clearly *this* is not intended: but, in order to exclude such a possibility, we must extend yet further the meaning of the phrase, and, if two Lines be said to have 'the same direction,' we must understand 'pairs of directions which can be arranged so as to be the same'; and if they be said to have 'different directions,' we must understand 'pairs of directions which cannot be arranged so as to be the same.'

Nie. Yes, that expresses our meaning.

Min. You must admit, I think, that your theory of direction involves a good deal of obscurity at the very outset. However, we have cleared it up, and will not use the word 'opposite' again. Tell me now whether you accept this as a correct Definition of the phrases 'the same direction' and 'different directions,' when used of a Pair of infinite Lines which have a common point:—

If two infinite Lines, having a common point, coincide, they have 'the same direction'; if not, they have 'different directions.'

Nie. We accept it.

Min. And, since a finite Line has the same direction as the infinite Line of which it is a portion, we may generalise thus :—'Coincidental Lines have the same direction. Non-coincidental Lines, which have a common point, have different directions.'

But it must be carefully borne in mind that we have as yet no geometrical meaning for these phrases, *unless when applied to two Lines which have a common point.*

Nie. Allow me to remark that what *you* call 'coincidental Lines' *we* call 'the same Line' or 'parts of the same Line,' and that what *you* call 'non-coincidental Lines' *we* call 'different Lines.'

Min. I understand you : but I cannot employ these terms, for two reasons : first, that your phrase 'the same Line' loses sight of a fact I wish to keep in view, that we are considering a *Pair* of Lines ; secondly, that your phrase 'different Lines' might be used, with strict truth, of two different portions of the same infinite Line, so that it is not definite enough for my purpose.

Let us now proceed 'to consider the relations of two or more straight Lines in one Plane in respect of direction.'

And first let me ask which of the propositions of Table II you wish me to grant you as an axiom ?

Nie. (*proudly*) Not one of them ! We have got a new patent process, the 'direction' theory, which will dispense with them all.

Min. I am *very* curious to hear how you do it.

NIEMAND *reads.*

P. II. Ax. 6. 'Two different Lines may have either the same or different directions.'

Min. That contains two assertions, which we will consider separately. First, you say that 'two different Lines (i. e. 'non-coincidental Lines,' or 'Lines having a separate point') 'may have the same direction'. Now let us understand each other quite clearly. We will take a fixed Line to begin with, and a certain point on it: there is no doubt that we can draw, through that point, a second Line coinciding with the first: the direction of this Line will of course be 'the same' as the direction of the first Line; and it is equally obvious that if we draw the second Line in any *other* direction, so as *not* to coincide with the first, its direction will *not* be 'the same' as that of the first: that is, they will have 'different' directions. If we want a geometrical definition of the assertion that this second Line has 'the same direction' as the first Line, we may take the following:—'having such a direction as will cause the Lines to be the same Line.' If we want a geometrical construction for it, we may say 'take any other point on the fixed Line; join the two points, and produce the Line, so drawn, at both ends': this construction we know will produce a Line which will be 'the same' as the first Line, and whose direction will therefore be 'the same' as that of the first Line. If, in a certain diagram, whose geometrical history we know, we want to test whether two Lines, passing through a common point, have, or have not, 'the same direction,' we have simply to take any other point on one of the Lines, and observe whether the other Line does, or does not, pass through it. This relationship of direction, which *you* call 'having the same direction,' and *I* 'having identical directions,' we may express by the word 'co-directional.'

Nie. All very true. My only puzzle is, why you have explained it at such enormous length: my meerschaum has gone out while I have been listening to you!

Min. Allow me to hand you a light. As to the 'enormous length' of my explanation, we are in troubled waters, my friend! There are breakers ahead, and we cannot 'heave the lead' too often.

Nie. It is 'lead' indeed!

Min. Let us now return to our fixed Line: and this time we will take a point *not* on it, and through this point we will draw a second Line. You say that we can, if we choose, draw it in 'the same direction' as that of the first Line?

Nie. We do.

Min. In that case let me remind you of the warning I gave you a few minutes ago, that we have no geometrical meaning for the phrase 'the same direction,' *unless when used of Lines having a common point.* What geometrical meaning do you attach to the phrase when used of other Lines?

Nie. (*after a pause*) I fear we cannot give you a geometrical definition of it at present.

Min. No? Can you construct such Lines?

Nie. No, but really that is not necessary. We allow of 'hypothetical constructions' now-a-days.

Min. Well then, can you test whether a given Pair of Lines *have* this property? I mean, if I give you a certain diagram, and tell you its geometrical history, can you pronounce, on a certain Pair of finite Lines, which have no visible common point, as to whether they have this property?

Nie. We cannot undertake it.

Min. You ask me, then, to believe in the reality of a class of 'Pairs of Lines' possessing a property which you can neither define, nor construct, nor test?

Nie. We can do none of these things, we admit: but yet the class is not quite so indefinite as you think. We can give you a geometrical *description* of it.

Min. I shall be delighted to hear it.

Nie. We have agreed that a Pair of coincidental finite Lines have a certain relationship of direction, which we call 'the same direction,' and which you allow to be an intelligible geometrical relation?

Min. Certainly.

Nie. Well, all we assert of this new class is that their relationship of direction is identical with that which belongs to coincidental Lines.

Min. It cannot be identical in *all* respects, for it certainly differs in this, that we cannot reach the conception of it by the same route. I can form a conception of 'the same direction,' when the phrase is used of two Lines which have a common point, but it is only by considering that one 'falls on' the other—that they have all other points common—that they coincide. When you ask me to form a conception of this relationship of direction, when asserted of other Lines, you know that none of these considerations will help me, and you do not furnish me with any substitutes for them. To me the relationship does *not* seem to be identical: I should prefer saying that separational Lines have 'collateral,' or 'corresponding,' or 'separational' directions, to using the phrase 'the same direction' over again. It is, of course, true that 'collateral' directions produce the same results, as to angles made with a transversal, as 'identical' directions; but this seems to me to be a Theorem, not an Axiom.

Nie. You say that the relationship does *not* seem to you to be identical. I should like to know *where* you think you perceive any difference?

Min. I will try to make my meaning clearer by an illustration.

Suppose that I and several companions are walking along a railway, which will take us to a place we wish to visit. Some amuse themselves by walking on one of the rails; some on another; others wander along the line, crossing and recrossing. Now as we are all bound for the same place, we may say, roughly speaking, that we are *all* moving 'in the same direction': but that is speaking very roughly indeed. We make our language more exact, if we exclude the wanderers, and say that those who are walking along the rails are so moving. But it seems to me that our phrase becomes still more exact, if we limit it to those who are walking on one and the same rail.

As a second illustration, suppose two forces, acting on a certain body; and let them be equal in amount and opposite in direction. Now, if they are acting along the same

Line, we know that they neutralise each other, and that the body remains at rest. But if one be shifted ever so little to one side, so that they act along parallel Lines, then, though still equal in amount and (according to the 'direction' theory) opposite in direction, they no longer neutralise each other, but form a 'couple.'

As a third illustration, take two points on a certain Plane. We may, first, draw a Line through them and cause them to move along that Line: they are then undoubtedly moving 'in the same direction.' We may, secondly, draw two Lines through them, which meet or at least would meet if produced, and cause them to move along those Lines: they are then undoubtedly moving 'in different directions.' We may, thirdly, draw two parallel Lines through them, and cause them to move along those Lines. Surely this is a new relationship of motion, not absolutely identical with either of the former two? But if this new relationship be not absolutely identical with that named 'in the same direction,' it must belong to the class named 'in different directions.'

Still, though this new relationship of direction is not identical with the former in *all* respects, it is in *some*: only, to prove this, we must use *some* disputed Axiom, as it will take us into Table II. For instance, they are identical as to angles made with transversals: this fact is embodied in Tab. II. 4. (See p. 34). Would you like to adopt that as your Axiom?

Nie. No. We are trying to dispense with Table II altogether.

Min. It is a vain attempt.

There is another remark I wish to make, before considering your second assertion. In asserting that there is a real class of non-coincidental Lines that have 'the same direction,' are you not also asserting that there is a real class of Lines that have no common point? For, if they *had* a common point, they must have '*different* directions.'

Nie. I suppose we are.

Min. We will then, if you please, credit you with an Axiom you have not expressed, viz. 'It is possible for two Lines to have no common point.' And here I must express an opinion that this ought to be *proved*, not assumed. Euclid has proved it in I. 27, which rests on no disputed Axiom; and I think it may be recorded as a distinct defect in your treatise, that you have assumed, as axiomatic, a truth which Euclid has *proved*.

My conclusion, as to this first assertion of yours, is that it is most decidedly *not* axiomatic.

Let us now consider your second assertion, that some non-coincidental Lines have 'different directions.' Here I must ask, as before, are you speaking of Lines which have a common point? If so, I am quite ready to grant the assertion.

Nie. Not exactly. It is rather a difficult matter to explain. The Lines we refer to *would*, as a matter of fact, meet if produced, and yet we do not suppose that fact known in speaking of them. What we ask you to believe is that there is a real class of non-coincidental finite Lines, which we do not yet know to have a common point, but which have 'different directions.' We shall assert presently, in another Axiom, that such Lines will meet if produced; but we ask you to believe their reality independently of that fact.

Min. But the only geometrical meaning I know of, as yet, for the phrase 'different directions,' refers to Lines known to have a common point. What geometrical meaning do you attach to the phrase when used of other Lines?

Nie. We cannot define it.

Min. Nor construct it? Nor test it?

Nie. No.

Min. You ask me, then, to believe in the reality of *two* classes of 'pairs of Lines,' each possessing a property that you can neither define, nor construct, nor test?

Nie. That is true. But surely you admit the reality of the second class? Why, intersectional Lines are a case in point.

Min. Certainly. And so much I am willing to grant you. I allow that *some* non-coincidental Lines, viz. intersectional Lines, have 'different directions.' But as to 'the *same* direction,' you have given me no reason whatever for believing that there are *any* non-coincidental Lines which possess that property.

Nie. But surely there are two real distinct classes of non-coincidental Lines, 'intersectional' and 'separational'?

Min. Yes. Thanks to Euc. I. 27, you may now assume the reality of both.

Nie. And you will hardly assert that the relationship of direction, which belongs to a Pair of intersectional Lines, is identical with that which belongs to a Pair of separational Lines?

Min. I do not assert it.

Nie. And you allow that intersectional Lines have 'different directions'?

Min. Yes. Are you going to argue, from that, that separational Lines must have 'the *same* direction'? Why may I not say that intersectional Lines have *one* kind of 'different directions' and that separational Lines have *another* kind?

Nie. But *do* you say it?

Min. Certainly not. There is no evidence, at present, one way or the other. For anything we know, Pairs of separational Lines may always have 'the same direction,' or they may always have 'different directions,' or there may be Pairs of each kind. I fear I must decline to grant the first part of your Axiom altogether, and the second part in the sense of referring to Lines not known to have a common point. You may now proceed.

NIEMAND *reads.*

P. 11. Ax. 7. 'Two different straight Lines which meet one another have different directions.'

Min. That I grant you, heartily. It is, in fact, a Definition for the phrase 'different directions,' when used of Lines which have a common point.

NIEMAND *reads.*

P. 11. Ax. 8. 'Two straight Lines which have different directions would meet if prolonged indefinitely.'

Min. Am I to understand that, if we have before us a Pair of finite Lines which are not known to have a common point, but of which we *do* know that they have different directions, you ask me to believe that they will meet if produced?

Nie. That is our meaning.

Min. We had better heave the lead once more, and return to our fixed Line, and a point not on it, through which we wish to draw a second Line. You ask me to grant that, if it be drawn so as to have a direction 'different' from that of the first Line, it will meet it if prolonged indefinitely?

Nie. That is our humble petition.

Min. Will you be satisfied if I grant you that *some* Lines, so drawn, will meet the first Line? *That* I would grant you with pleasure. I could draw millions of Lines which would fulfil the conditions, by simply taking points at random on the given Line, and joining them to the given point. Every Line, so constructed, would have a direction 'different' from that of the given Line, and would also meet it.

Nie. We will *not* be satisfied, even with millions! We ask you to grant that *every* Line, drawn through the given point with a direction 'different' from that of the given Line, will meet the given Line: and we ask you to grant this independently of, and antecedently to, any other information about the Lines except the fact that they have 'different' directions.

Min. But what meaning am I to attach to the phrase 'different directions,' independently of, and antecedently to, the fact that they have a common point?

Nie. (*after a long silence*) I fear we can suggest none.

Min. Then I must decline to accept the Axiom.

Nie. And yet this Axiom is the converse of the preceding, which you granted so readily.

Min. The *technical* converse, my good sir, not the *logical*! I will not suspect you of so gross a logical blunder as the attempt to convert a universal affirmative *simpliciter* instead of *per accidens.* The only converse, as you are no doubt aware, to which you have any *logical* right, is '*Some* Lines, which have "different directions," would meet if produced'; and *that* I grant you. It is true of intersectional Lines, and I would limit the Proposition *to* such Lines, so that it would be equivalent to 'Lines, which would meet if produced, would meet if produced'—an indisputable truth, but *not* remarkable for novelty! You may proceed.

Nie. I beg to hand in this diagram, and will read you our explanation of it :—

A B C D E F

'Thus *A* and *B* in the figure have the same direction; and *C* and *D*, which meet, have different directions; and *E* and *F*, which have different directions, would meet if produced far enough.'

Min. I grant the assertion about *C* and *D*; but I am wholly unable to guess on what grounds you expect me to grant that *A* and *B* 'have the same direction,' and that *E* and *F* 'have different directions.' Do you expect me to judge by eye? How if the lines were several yards apart? Is *this* what Geometry is coming to? Proceed.

Niemand *reads*.

Def. 19. 'Straight Lines, which are not parts of the same straight Line, but have the same direction, are called *Parallels*.'

Min. A *Definition* is of course unobjectionable, since it does not assert the *existence* of the thing defined: in fact, it *asserts* nothing except the meaning which you intend to attach to the word 'parallel.' But, as this word is used in different senses, I will thank you to substitute for it, in what you have yet to say about this matter, the phrase 'having a separate point, but the same direction,' which you may condense into one compound word, if you like:—'sepuncto-codirectional.'

Nie. (*sighing*) A terrible word! And I shall have to use it so often!

Min. I will try to abridge it for you. Let us take 'sep-' and 'cod-' from the beginnings of the two words, and '-al' for a termination. That will give us 'sepcodal.'

Nie. That sounds a little harsh.

Min. 'What? Is it harder, Sirs, than Gordon,
Colkitto, or Macdonald, or Galasp?'

Nie. (*doubtfully*) I *think* I prefer it to Colkitto.' But it is from you Moderns I have learned to be so sensitive about long words. How I would have liked to take you to an Egyptian restaurant I used to frequent, centuries ago, in a phantasmic sort of way, if only to hear the names of some of the dishes! Why, one thought nothing of seeing a gentleman rush in, carpet-bag in hand, and shout out 'ὑᾷτερ̨!' (that was the way we addressed the attendant in those days) ‘A plate of λεπαδοτεμαχοσελαχογαλεοκρανιολειψανοδριμυποτριμματοσιλφιοπαραομελιτοκατακεχυμενοκιχλεπικοσσυφοφαττοπεριστεραλεκτρυονοπτεγκεφαλοκιγκλοπελειολαγωοσιραιοβαφητραγανοπτερύγων, and look sharp about it! I'm in a hurry!’

Min. If the gentleman wanted to catch his train—by the way, *had* they trains in Egypt in ancient days?

Nie. Certainly. Read your ‘Antony and Cleopatra,’ Act I, Scene 1. ‘*Exeunt Antony and Cleopatra with their train.*’

Min. In that case, wouldn't it be enough to say ‘A plate of λεπαδο’?

Nie. Most certainly *not*—at least not in a *fashionable* restaurant. But this is a digression. I am willing to adopt the word ‘sepcodal.’

Min. Now, before you read any more, let us get a clear idea of your Definition. We know of two real classes of Pairs of Lines, ‘coincidental’ and ‘intersectional’; and to these we may (if we credit you with a Corollary to Euc. I. 27, ‘It is possible for two Lines to have no common point’) add a third class, which we may call ‘separational.’

We also know that if a Pair of Lines has a common point, and no separate point, it belongs to the first class; if a common point, and a separate point, to the second. Hence all Pairs of Lines, having a common point, must belong to one or other of these classes. And since a Pair, which has *no* common point, belongs to the third class, we see that *every* conceivable Pair of Lines must belong to one of these three classes.

We also know that—

Nie. (*sighing deeply*) You are heaving the lead again!

Min. I am: but we shall be in calmer water soon.

We also know that the 'Coincidental' class possesses two properties—they are coincidental and have identical directions; and that the 'Intersectional' class also possesses two properties—they are intersectional and have different directions.

Now if you choose to frame a Definition by denying one property of each of these two classes, any Pair of Lines, so defined, is excluded from both of these classes, and must, *if it exist at all*, belong to the 'Separational' class. Remember, however, that you *may* have so framed your Definition as to exclude your Pair of Lines from *existence*. For instance, if you choose to combine two contradictory conditions of direction, and to say that Lines, which have identical *and* intersectional directions, are to be called so-and-so, you are simply describing a nonentity.

Nie. That is all quite clear.

Min. Your Definition, then, amounts to this:—Lines, which are *not* coincidental, but which have identical directions, are said to be 'sepcodal.'

Nie. It does.

Min. Well, here is another Definition for Parallels, which will answer your purpose just as well:—'Lines, which are *not* intersectional, but which have different directions.'

Nie. But I think I can prove to you that you have now done the very thing you cautioned me against: you have annihilated your Pair of Lines.

Min. That is a matter which we need not consider at present. Proceed.

NIEMAND *reads.*

P. 11. 'From this Definition, and the Axioms above given, the following results are immediately deduced:

(1) That parallel—I beg your pardon—that 'sepcodal' Lines would not meet however far they were produced. For if they met——'

Min. You need not trouble yourself to prove it. I grant that, *if* such Lines existed, they would not meet. Your assertion is simply the Contranominal of Ax. 7 (p. 115), and therefore is necessarily true if the subject be real.

But remember that, though I have granted to you that, if we are given a Line and a point not on it, we can draw, through the point, *a certain* Line separational from the given Line, we do not yet know that it is *the only* such Line. *That* would take us into Table II. With our present knowledge, we must allow for the possibility of drawing any number of Lines through the given point, all separational from the given Line: and all I grant you is, that your ideal 'sepcodal' Line will, *if it exist at all,* be one of this group.

NIEMAND *reads.*

(2) 'That Lines which are sepcodal with the same Line are sepcodal with each other. For——'

Min. Wait a moment. I observe that you say that such Lines are sepcodal with each other. Might they not be '*com*puncto-codirectional'?

Nie. Certainly they might: but we do not wish to include that case in our predicate.

Min. Then you must limit your subject, and say '*different* Lines.'

Nie. Very well.

Reads.

'That different Lines, which are sepcodal with the same Line, are sepcodal with each other. For they each have the same direction as that Line, and therefore the same direction as the other.'

Min. I am willing to grant you, without any proof, that, *if* such Lines existed, they would have the same direction with regard to each other. The phrase '*they each have*' is not remarkably good English. However, you may proceed.

NIEMAND *reads.*

P. 12. Ax. 9. 'An angle may be conceived as transferred from one position to another, the direction of its arms remaining the same.'

Min. Let us first consider the right arm by itself. You assert that it may be transferred to a new position, its direction remaining the same?

Nie. We do.

Min. You might, in fact, have here inserted an Axiom 'A Line may be conceived as transferred from one position to another, its direction remaining the same'?

Nie. That would express our meaning.

Min. And this is virtually identical with your Axiom 'Two different Lines may have the same direction'?

Nie. Certainly. They embody the same truth. But the one contemplates a single Line in two positions, and the other contemplates two Lines: the difference is very slight.

Min. Exactly so. Now let me ask you, do you mean, by the word 'angle,' a constant or a variable angle?

Nie. I do not quite understand your question.

Min. I will put it more fully. Do you mean that the arms of the angle are rigidly connected, so that it cannot change its magnitude, or that they are merely hinged loosely together, as it were, so that it depends entirely on the relative motions of the two arms whether the angle changes its magnitude or not?

Nie. Why are we bound to settle the question at all?

Min. I will tell you why. Suppose we say that the arms are merely hinged together: in that case all you assert is that each arm may be transferred, its direction remaining the same; that is, you merely assert your 6th Axiom twice over, once for the right arm and once for the left arm; and you do *not* assert that the angle will retain its magnitude. But in the Theorem which follows, you clearly regard it as a constant angle, for you say 'the angle AOD would coincide with the angle EKH. *Therefore* the angle $AOD = EKH$.' But the 'therefore' would have no force if AOD could change its magnitude. Thus you would be deducing, from an Axiom where 'angle' is used in a peculiar sense, a conclusion in which it bears its ordinary sense. You have heard of the fallacy '*A dicto secundum Quid ad dictum Simpliciter*'?

Nie. (*hastily*) We are not going to commit ourselves to *that*. You may assume that we mean, by 'angle,' a rigid angle, which cannot change its magnitude.

Min. In that case you assert that, when a pair of Lines, terminated at a point, is transferred so that its vertex has a new position, these three conditions can be simultaneously fulfilled :—

(1) the right arm has 'the same direction' as before;

(2) the left arm has 'the same direction' as before;

(3) the magnitude of the angle is unchanged.

Nie. We do not dispute it.

Min. But any *two* of these conditions are sufficient, without the third, to determine the new state of things. For instance, taking (1) and (3), if we fix the position of the right arm, by giving it 'the same direction' as before, and also keep the magnitude of the angle unchanged, is not that enough to fix the position of the left arm, without mentioning (2)?

Nie. It certainly is.

Min. Your Axiom asserts, then, that any two of these conditions lead to the third as a necessary result?

Nie. It does.

Min. Your Axiom then contains *two* distinct assertions: the data of the first being (1) and (3) [or (2) and (3), which lead to a similar result], the data of the second being (1) and (2). These I will state as two separate Axioms :—

9 (α). If a Pair of Lines, terminated at a point, be transferred to a new position, so that the direction of one of the Lines, and the magnitude of the included angle, remain the same; the direction of the other Line will remain the same.

9 (β). If a Pair of Lines, terminated at a point, be transferred to a new position, so that their directions remain

the same; the magnitude of the included angle will remain the same.

Have I represented your meaning correctly?

Nie. We have no objection to make.

Min. We will return to this subject directly. I must now ask you to read the enunciation of Th. 4, omitting, for simplicity's sake, all about supplementary angles, and assuming the Lines to be taken 'the same way.'

NIEMAND *reads.*

P. 12. Th. 4. 'If two Lines are respectively sepcodal with two other Lines, the angle made by the first Pair will be equal to the angle made by the second Pair.'

Min. The 'sep' is of course superfluous, for if the Lines are '*com*puncto-codirectional,' it is equally true. May I re-word it thus?—

'If two Pairs of Lines, each terminated at a point, be such that the directions of one Pair are respectively the same as those of the other; the included angles are equal.'

Nie. Yes, if you like.

Min. But surely the only difference, between Ax. 9 (β) and this, is that in the Axiom we contemplated a single Pair of Lines transferred, while *here* we contemplate two Pairs?

Nie. That is the only difference, we admit.

Min. Then I must say that it is anything but good logic to take two Propositions, distinguished only by a trivial difference in form, and to call one an Axiom and the other a Theorem deduced from it! A very gross case of '*Petitio Principii,*' I fear!

Nie. (*after a long pause*) Well! We admit that it is *not* exactly a Theorem: it is only a new form of the Axiom.

Min. Quite so: and as it is a more convenient form for my purpose, I will with your permission adopt it as a substitute for the Axiom. Now as to the corollary of this Theorem: *that*, I think, is merely a particular case of Ax. 9 (β), one of the arms being slid along the infinite Line of which it forms a part, and thus of course having 'the same direction' as before?

Nie. It is so.

Min. And, as this is a more convenient form still, I will restate your assertions, limiting them to this particular case:—

Ax. 9 (a). Lines, which make equal corresponding angles with a certain transversal, have the same direction.

Ax. 9 (β). Lines, which have the same direction, make equal corresponding angles with any transversal.

Am I right in saying that these two assertions are virtually involved in your Axiom?

Nie. We cannot deny it.

Min. Now in 9 (a) you ask me to believe that Lines possessing a certain geometrical property, which can be defined, constructed, and tested, possess also a property which, in the case of different Lines, we can neither define, nor construct, nor test. There is nothing axiomatic in this. It is much more like a Definition of 'codirectional' when asserted of different Lines, for which we have as yet no Definition at all. Will you not permit me to insert it, as a Definition, before Ax. 6 (p. 108)? We might word it thus:—

'If two different Lines make equal angles with a certain transversal, they are said to have the same direction: if unequal, different directions.'

This interpolation would have the advantage of making Ax. 6 (which I have hitherto declined to grant) indisputably true.

Nie. (*after a pause*) No. We cannot adopt it as a Definition so early in the subject.

Min. You are right. You probably saw the pitfall which I had ready for you, that this same Definition would make your 8th Axiom (p. 115) exactly equivalent to Euclid's 12th! From this catastrophe you have hitherto been saved solely by the absence of geometrical meaning in your phrase 'the same direction,' when applied to different Lines. Once define it, and you are lost!

Nie. We are aware of that, and prefer all the inconvenience which results from the absence of a Definition.

Min. The 'inconvenience,' so far, has consisted of the ruin of Ax. 6 and Ax. 8. Let us now return to Ax. 9.

As to 9 (β), it is of course obviously true with regard to *coincidental* Lines: with regard to *different* 'Lines, which have the same direction,' I grant you that, *if* such Lines existed, they *would* make equal corresponding angles with any transversal; for they would then have a relationship of direction identical with that which belongs to coincidental Lines. But all this rests on an 'if'—*if* they existed.

Now let us combine 9 (β) with Axiom 6, and see what it is you ask me to grant. It is as follows:—

'There can be a Pair of different Lines that make equal angles with *any* transversal.'

I am not misrepresenting you, I think, if I say that you propound this as axiomatic truth—which, I need hardly remark, is a corollary deducible from the fourth Proposition in Table II. (see p. 34).

Nie. We accept the responsibility of the two Axioms separately, but not of a logical deduction from the two.

Min. There are certainly *some* logical deductions from Axioms (Contranominals for instance) that are not so axiomatic as the Axioms from which they come: but surely if you tell me 'it is axiomatic that *X* is *Y*' and 'it is axiomatic that *Y* is *Z*,' it is much the same as saying 'it is axiomatic that *X* is *Z*'?

Nie. It is very like it, we admit.

Min. Now take one more combination. Take 9 (α) and 9 (β). We thus eliminate the mysterious property altogether, and get a Proposition whose subject and predicate are perfectly definite geometrical conceptions—a Proposition which you assert to be, if not perfectly axiomatic, yet so nearly so as to be easily deducible from two Axioms—a Proposition which again lands us in Table II, and which, I will venture to say, is less axiomatic than any Proposition in that Table that has yet been proposed as an Axiom. We get *this*:—

'Lines which make equal corresponding angles with a certain transversal do so with *any* transversal,' which is Tab. II. 4 (see p. 34).

Here we have, condensed into one appalling sentence, the whole substance of Euclid I. 27, 28, and 29 (for the fact that the lines are 'separational' may be regarded as merely a go-between). Here we have the whole difficulty of Parallels swallowed at one gulp. Why, Euclid's much-abused 12th Axiom is nothing to it! If we had (what I fear has yet to be discovered) a unit of 'axiomaticity,' I should expect to find that Euclid's 12th Axiom (which you call in your Preface, at p. xiii, 'not axiomatic') was twenty or thirty times as axiomatic as this! I need not ask you for any further proof of Euc. I. 32. This wondrous Axiom, or quasi-Axiom, is quite sufficient machinery for your purpose, along with Euc. I. 13, which of course we grant you. Have you thought it necessary to provide any other machinery?

Nie. No.

Min. Euclid requires, besides I. 13, the following machinery:—Props. 4, 5, 7, 8, 15, 16, Ax. 12, Props. 27, 28, and 29. And for all this you offer, as a sufficient substitute, one single Axiom!

Nie. *Two*, if you please. You are forgetting Ax. 6.

Min. No, I repeat it—one single Axiom. Ax. 6 is contained in Ax. 9 (*a*)·: when the subject is known to be real, the Proposition necessarily asserts the reality of the predicate.

Nie. That we must admit to be true.

Min. I need hardly say that I must decline to grant this so-called 'Axiom,' even though its collapse should involve that of your entire system of 'Parallels.' And now that we have fully discussed the subject of direction, I wish to ask you one question which will, I think, sum up the whole difficulty in a few words. It is, in fact, *the* crucial test as to whether 'direction' is, or is not, a logical method of proving the properties of Parallels.

You assert, as axiomatic, that different Lines exist, whose relationship of direction is identical with that which exists between coincidental Lines.

Nie. Yes.

Min. Now, does the phrase 'the same direction,' when used of two Lines not known to have a common point, convey to your mind a clear geometrical conception?

Nie. Yes, we can form a clear idea of it, though we cannot define it.

Min. And is that idea (this is the crucial question) *independent of all subsequent knowledge of the properties of Parallels?*

Nie. We believe so.

Min. Let us make sure that there is no self-deception in this. You feel certain you are not unconsciously picturing the Lines to yourself as being equidistant, for instance?

Nie. No, they suggest no such idea to us. We introduce the idea of equidistance later on in the book, but we do not feel that our first conception of 'the same direction' includes it at all.

Min. I think you are right, though Mr. Cuthbertson, in his 'Euclidian Geometry,' says (Pref. p. vi.) 'the conception of a parallelogram is not that of a figure whose opposite sides will never meet . . . . , but rather that of a figure whose opposite sides are equidistant.' But do you feel equally certain that you are not unconsciously using your subsequent knowledge that Lines exist which make equal angles with all transversals?

Nie. We are not so clear about *that.* It is, of course, extremely difficult to divest one's mind of all later knowledge, and to place oneself in the mental attitude of one who is totally ignorant of the subject.

Min. Very difficult, no doubt, but absolutely essential, if you mean to write a book adapted to the use of beginners. My own belief as to the course of thought needed to grasp the theory of 'direction' is this:—first you grasp the idea of 'the same direction' as regards Lines which have a common point; next, you convince yourself, by some *other* means, that different Lines exist which make equal angles with all transversals; thirdly, you go back, armed with this new piece of knowledge, and use it unconsciously, in forming an idea of 'the same direction' as regards different Lines. And I believe that the course of thought in the mind of a beginner is simply this:—he grasps, easily enough, the idea of 'the same direction' as regards Lines which have a common point; but when you put before him the idea of *different* Lines, and ask him to realise the meaning of the phrase, when applied to such Lines, he, finding that the former geometrical conception of 'coincidence' is not applicable in this case, and knowing nothing of the idea, which is latent in *your* mind, of Lines which make equal angles with all transversals, simply fails to attach *any* idea at all to the phrase, and accepts it blindly, from faith in his teacher, and is from that moment, until he reaches the Theorem about transversals, walking in the dark.

Nie. If this be true, of course the theory of 'direction,' however beautiful in itself, is not adapted for purposes of teaching.

Min. That is my own firm conviction. But I fear I may have wearied you by discussing this matter at such great length. Let us turn to another subject. What is your practical test for knowing whether two finite Lines will meet if produced?

Nie. You have already heard our 8th Axiom (p. 11). 'Two straight lines which have different directions would meet if produced.'

Min. But, even if that were axiomatic (which I deny), it would be no *practical* test, for you have admitted that you have no means of knowing whether two Lines, not known to have a common point, have or have not different directions.

Nie. We must refer you to p. 14. Th. 5. Cor. 2, where we prove that Lines, which make equal angles with a certain transversal, have the same direction.

Min. Which you had already asserted, if you remember, in Ax. 9.

Nie. Well then, we refer you to Ax. 9 as containing the same truth.

Min. And having got that truth, whether lawfully or not, what do you do with it?

Nie. Why, surely it is almost the same as saying that, if they make *un*equal angles, they have *different* directions.

Min. And what then?

Nie. Then, combining this with the Axiom you refused to grant, namely, that Lines having different directions will meet, we get a practical test, such as you were asking for.

Min. (*dreamily*) I see! You get rid of the 'different directions' altogether, and the result is that 'Lines, which make unequal angles with a certain transversal, will meet if produced,' which is Tab. II. 2 (see p. 34). And this you assert as axiomatic truth?

Nie. (*uneasily*) Yes.

Min. Surely I have read something like it before? Could it have been Euclid's 12th Axiom? And have I not somewhere read words like these:—'Euclid's treatment of parallels distinctly breaks down in Logic. It rests on an Axiom which is not axiomatic'?

Nie. We have nowhere *stated* this Axiom which you put into our mouth.

Min. No? Then how, may I ask, do you prove that particular Lines *will* meet? You *must* have to prove it sometimes, you know.

Nie. We have not had to prove it anywhere, that we are aware of.

Min. Then there must be some gaps in your arguments. Let us see. Please to turn to p. 46. Prob. 7. Here you make, at the ends of a Line *CD*, angles equal to two given angles (which, as you tell us below, 'must be together less than two right angles'), and you then say 'let their sides meet in *O*.' How do you know that they *will* meet?

Nie. You have found *one* hiatus, we grant. Can you point out another in the whole book?

Min. I can. At p. 70 I find the words 'Join *QG*, and produce it to meet *FH* produced in *S*.' And again at p. 88. 'Hence the centre must be at *O*, the point of intersection of these perpendiculars.' In both these cases I would ask, as before, how do you know that the Lines in question *will* meet?

Nie. We had not observed the omissions before, and we must admit that they constitute a serious hiatus.

Min. A most serious one. A student, who had been

taught such proofs as these, would be almost sure to try the plan in cases where the Lines would *not* really meet, and his assumption would lead him to results more remarkable for novelty than truth.

Let us now take a general survey of your book. And First, as to the Propositions of Euclid which you omit—

Nie. You are alluding to Prop. 7, I suppose. Surely its only use is to prove Prop. 8, which we have done very well without it.

Min. That is quite a venial omission. The others that I miss are 27, 28, 29, 30, 33, 34, 35, 36, 37, 38, 39, 40, 41, and 43: rather a formidable list.

Nie. You are much mistaken! Nearly all of those are in our book, or could be deduced in a moment from theorems in it.

Min. Let us take I. 34 as an instance.

Nie. *That* we give you, almost in the words of Euclid, at p. 37.

Reads.

Th. 22. 'The opposite angles and sides of a Parallelogram will be equal, and the diagonal, or the Line which joins its opposite angles, will bisect it.'

Min. Well, but *your* Parallelogram is not what Euclid contemplates. *He* means by the word that the opposite sides are separational—a property whose reality he has demonstrated in I. 27; whereas *you* mean that they have the same direction—a property whose reality, when asserted of different Lines, has nowhere been satisfactorily proved.

Nie. We have proved it at p. 14. Th. 5. Cor. 2.

Min. Which, if traced back, will be seen to depend ultimately on your 6th Axiom, where you *assume* the reality of such Lines. But, even if your Theorem *had* been shown to refer to a real figure, how would that prove Euc. I. 34?

Nie. You only need the link that separational Lines have the same direction.

Min. Have you supplied that link?

Nie. No: but the reader can easily make it for himself. It is the 'Contranominal' (as you call it) of our 8th Axiom, 'two straight Lines which have different directions would meet if prolonged indefinitely.'

Min. Your pupils must be remarkably clever at drawing deductions and filling up gaps in an argument, if they usually supply that link, as well as the proof that separational Lines exist at all, for themselves. But, as *you* do not supply these things, it seems fair to say that your book omits all the Propositions which I have enumerated.

I will now take a general survey of your book, and select a few points which seem to call for remark.

MINOS *reads.*

P. 14. Th. 5. Cor. 1. 'Hence if two straight Lines which are not parallel are intersected by a third, the alternate angles will be not equal, and the interior angles on the same side of the intersecting Line will be not supplementary.' Excuse the apparent incivility of the remark, but this Corollary is false.

Nie. You amaze me!

Min. You have simply to take, as an instance, a Pair of *coincidental* Lines, which most certainly answer to your description of 'not parallel.'

Nie. It is an oversight.

Min. So I suppose: it is a species of literary phenomenon in which your Manual is rich.

Your proof of Cor. 2. is a delicious collection of negatives.

Reads.

'Cor. 2. *Hence also if the corresponding angles are equal, or the alternate angles equal, or the interior angles supplementary, the Lines will be parallel.*

'For they can*not* be *not* parallel, for then the corresponding and alternate angles would be *un*equal by Cor. 1.'

Should I be justified in calling this a somewhat *knotty* passage?

Nie. You have no right to make such a remark. It is a mere jest!

Min. Well, we will be serious again.

At p. 9, you stated *more* than the data authorised: we now come to a set-off against this, since we shall find you asserting *less* than you ought to do. I will read the passage:—

P. 26. Th. 15. 'If two Triangles are equiangular to one another and have a side of the one equal to the corresponding side of the other, the Triangles will be equal in all respects.'

This contains a superfluous *datum*: it would have been

enough to say 'if two Triangles have two angles of the one equal to two angles of the other &c.'

Nie. Well, it is at worst a superfluity: the enunciation is really identical with Euclid's.

Min. By no means. The logical effect of a superfluous *datum* is to *limit* the extent of a Proposition: and, if the Proposition be 'universal,' it reduces it to 'particular'; i.e. it changes 'all *A* is *B*' into 'some *A* is *B*.' For suppose we take the Proposition 'all *A* is *B*,' and substitute for it 'all that is both *A* and *X* is *B*,' we may be *accidentally* making an assertion of the same extent as before, for it may happen that the whole class '*A*' possesses the property '*X*'; but, so far as logical *form* is concerned, we have reduced the Proposition to '*some* things that are *A* (viz. those which are also *X*) are *B*.'

I turn now to p. 27, where I observe a new proof for Euc. I. 24.

Nie. New and, we hope, neat and short.

Min. Charmingly neat and short, *as it stands*: but this method really requires the discussion of *five* cases, each with its own figure.

Nie. How do you make that out?

Min. The five cases are:—

(1) Vertical angles together less than two right angles, and adjacent base angles acute (the case you give).

(2) Adjacent base-angles right.

(3) Adjacent base-angles obtuse.

These two cases are proved along with the first.

(4) Vertical angles together equal to two right angles.

This requires a new proof, as we must substitute for the

words 'the bisector of the angle *FAC*,' the words 'the perpendicular to *FC* drawn through *A*.'

(5) Vertical angles together greater than two right angles.

This also requires a new proof, as we must insert, after the words 'the bisector of the angle *FAC*,' the words 'produced through *A*,' and must then prove (by your Th. 1) that the angles *OAC*, *OAF*, are equal.

On the whole, I take this to be the most cumbrous proof yet suggested for this Theorem.

We now come to what is probably the most extraordinary Corollary ever yet propounded in a geometrical treatise. Turn to pages 30 and 31.

Th. 20. 'If two triangles have two sides of the one equal to two sides of the other, and the angle opposite that which is not the less of the two sides of the one equal to the corresponding angle of the other, the triangles shall be equal in all respects.

'Cor. 1. If the side opposite the given angle were less than the side adjacent, there would be two triangles, as in the figure; and the proof given above is inapplicable.

'This is called *the ambiguous* case.'

The whole Proposition is a grand specimen of obscure writing and bad English, 'is' and 'are,' 'could,' and 'would,' alternating throughout with the most charming impartiality: but what impresses me most is the probable effect of this wondrous Corollary on the brain of a simple reader, coming breathless and exhausted from a death-struggle with the preceding theorem. I can imagine him saying wildly to himself 'If two Triangles fulfil such and such conditions, such and such things follow: but, if one of the conditions were to fail, *there would be two Triangles!* I must be dreaming! Let me dip my head in cold water, and read it all again. If two Triangles . . . there *would* be two Triangles. Oh, my poor brains!'

Nie. You are pleased to be satirical: it *is* rather obscure writing, we confess.

Min. It is indeed! You do well in calling it *the ambiguous case.*

At p. 33, I see the heading 'Theorems of equality': but you only give *two* of them, the second being 'the bisectors of the three angles of a triangle meet in one point,' which, as a specimen of 'Theorems of equality,' is probably unique in the literature of Geometry. I cannot wonder at your not attempting to extend the collection.

At p. 40 I read, 'It is assumed here that if a circle has one point inside another circle, the circumferences will intersect one another.' This I believe to be the boldest assumption yet made in Modern Geometry.

At pp. 40, 42 you assume a length 'greater than half' a given Line, without having shewn how to bisect Lines. Two cases of '*Petitio Principii.*' (See p. 58.)

P. 69. Here we have a Problem (which you call 'the quadrature of a rectilineal area') occupying three pages and a half. It is 'approached' by four 'stages,' which is a euphemism for saying that this fearful Proposition contains *four* of Euclid's Problems, viz. I. 42, 44, 45, and II. 14.

P. 73. 2. 'Find a point equally distant from three given

straight lines.' Is it fair to give this without any limitation? What if the given lines were parallel?

P. 84. 'If *A*, *B*, *C* . . . as conditions involve *D* as a result, and the failure of *C* involves a failure of *D*; then *A*, *B*, *D* . . . as conditions involve *C* as a result.' If not-*C* proves not-*D*, then *D* proves *C*. *A* and *B* are irrelevant and obscure the statement. I observe, in passing, the subtle distinction which you suggest between '*the* failure of *C*' and '*a* failure of *D*.' *D* is a habitual bankrupt, who has often passed through the court, and is well used to failures: but, when *C* fails, his collapse is final, and 'leaves not a wrack behind'!

P. 90. 'Given a curve, to ascertain whether it is an arc of a circle or not.' What does 'given a curve' mean? If it means a line drawn with ink on paper, we may safely say at once 'it is *not* a circle.'

P. 96. Def. 15. 'When one of the points in which a secant cuts a circle is made to move up to, and ultimately coincide with, the other, the ultimate position of the secant is called *the tangent* at that point.' (The idea of *the position of a Line* being itself a Line is queer enough: I suppose you would say '*the ultimate position* of Whittington was the Lord Mayor of London.' But this is by the way: of course you mean 'the secant in its ultimate position.') Now let us take three points on a circle, the middle one fixed, the others movable; and through the middle one let us draw two secants, each passing through one of the other points; and then let us make the other points 'move up to, and ultimately coincide with,' the middle one. We have no ground for saying that these two secants, in their ultimate positions, will coincide. Hence the phrase '*the* tangent' assumes, without proof, Th. 7. Cor. 1, viz. 'there can be only one tangent to a circle at a given point.' This is a '*Petitio Principii.*'

P. 97. Th. 6. The secant consists of two portions, each terminated at the fixed point. All that you prove here is that the portion which has hitherto cut the circle is ultimately outside: and you jump, without a shadow of proof, to the conclusion that the same thing is true of the *other* portion! Why should not the second portion begin to cut the circle at the precise moment when the first ceases to do so? This is another '*Petitio Principii.*'

P. 129, line 3 from end. 'Abstract quantities are the means that we use to express the concrete.' Excluding such physical 'means' as pen and ink or the human voice (to which you do not seem to allude), I presume that the 'means' referred to in this mysterious sentence are 'pure numbers.' At any rate the only instances given are 'seven, five, three.' Now take P. 130, l. 5, '*Abstract quantities* and *ratios* are precisely the same things.' Hence all ratios are numbers. But in the middle of the same page we read that 'all numbers are ratios, but all ratios are *not* numbers.' I leave this without further remark.

I will now sum up the conclusions I have come to with respect to your Manual.

(1) As to 'straight Lines' you suggest a useful extension of Euclid's Axiom.

(2) As to angles and right angles, your extension of

the limit of size is, in my opinion, objectionable. In other respects your language, though hazy, agrees on the whole with Euclid.

(3) As to 'Parallels,' there is a good deal to be said, and that not very flattering, I fear.

In Ax. 6, you assert the reality of different Lines having the same direction—a property you can neither define, nor construct, nor test.

You also assert (by implication) the reality of separational Lines, which Euclid *proves*.

You also assert the reality of Lines, not known to have a common point, but having different directions—a property you can neither define, nor construct, nor test.

In Ax. 8, you assert that the undefined Lines last mentioned would meet if produced.

These Axioms, therefore, are not axiomatic.

In proving result (2), you are guilty of the fallacy '*Petitio Principii*.'

In Ax. 9 and Th. 4 taken together, if the word 'angle' in Ax. 9 means 'variable angle,' you are guilty of the fallacy '*A dicto secundum Quid ad dictum Simpliciter*'; if 'constant angle,' of the fallacy '*Petitio Principii*.'

In Ax. 9 (α), you assert that Lines possessing a certain real geometrical property, viz. making equal angles with a certain transversal, possess also the before-mentioned undefined property. This is not axiomatic.

In Ax. 9 (β) combined with Ax. 6, you assert the reality of Lines which make equal angles with all transversals. This is not more axiomatic than Euc. Ax. 12.

In Ax. 9 (α) combined with Ax. 9 (β), you assert that Lines, which make equal angles with a certain transversal, do so with all transversals. This I believe to be the most unaxiomatic Axiom ever yet proposed.

(4) You furnish no practical test for the meeting of finite Lines, and consequently you never prove (however necessary for the matter in hand) that any particular Lines *will* meet. And when we come to examine what practical test can possibly be extracted from your Axioms, the only result is an imperfect edition of Euclid's 12th Axiom!

The sum total of the chief defects which I have noticed is as follows:—

fourteen of Euclid's Theorems in Book I. omitted;
seven unaxiomatic Axioms;
six instances of '*Petitio Principii.*'

The abundant specimens of logical inaccuracy, and of loose writing generally, which I have here collected would, I feel sure, in a mere popular treatise be discreditable—in a scientific treatise, however modestly put forth, deplorable—but in a treatise avowedly put forth as a model of logical precision, and *intended to supersede Euclid*, they are simply monstrous.

My ultimate conclusion on your Manual is that it has *no claim whatever* to be adopted as *the* Manual for purposes of teaching and examination.

ACT II.

SCENE VI.

§ 2. PIERCE.

'dum brevis esse laboro,
Obscurus fio.'

Nie. I lay before you '*An Elementary Treatise on lane and Solid Geometry*' by BENJAMIN PIERCE, A.M., Perkins Professor of Astronomy and Mathematics in Harvard University, published in 1872.

Min. As I have already considered, at great length, the subject of direction as treated by Mr. Wilson, I need not trouble you as to any matters where Mr. Pierce's treatment does not materially differ from his. Is there any material difference in the treatment of a straight line?

Nie. He has a Definition of direction which will, I think, be new to you:—

Reads.

P. 5. § 11, Def. 'The *Direction of a Line* in any part is the direction of a point at that part from the next preceding point of the Line.'

Min. That sounds mysterious. Which way along a Line are 'preceding' points to be found?

Nie. Both ways. He adds, directly afterwards, 'a Line has two different directions,' etc.

Min. So your Definition needs a postscript? That is rather clumsy writing. But there is yet another difficulty. How far from a point is the 'next' point?

Nie. At an infinitely small distance, of course. You will find the matter fully discussed in any work on the Infinitesimal Calculus.

Min. A most satisfactory answer for a teacher to make to a pupil just beginning Geometry! I see nothing else to remark on in your treatment of the Line, except that you state, as an Axiom, that 'a straight Line is the shortest way from one point to another.' I have already given, in my review of M. Legendre, my reasons for thinking that this is not a fair Axiom, and ought to be a Theorem (see p. 55).

There is nothing particular to notice in your treatment of angles and right angles. Let us go on to Parallels. How do you prove Euc. I. 32?

NIEMAND *reads.*

P. 9. § 27, Def. '*Parallel* Lines are straight Lines which have the same Direction.'

Min. I presume you do not mean to include coincidental Lines?

Nie. Certainly not. We see the omission. Allow us to insert the word 'different.'

Min. Very well. Then your Definition combines the two

properties 'different' and 'having the same direction.' Bear in mind that you have yet to prove the *reality* of such Lines. And may I request you in future to call such Lines 'sepcodal'? But if you wish to assert any thing of them which is also true of coincidental Lines, you had better drop the 'sep-' and simply call them 'Lines which have the same direction,' so as to include both classes.

Nie. Very well.

NIEMAND *reads.*

P. 9. § 28. Th. 'Sepcodal Lines cannot meet, however far they are produced.'

Min. Or rather '*could* not meet, if they existed.' Proceed.

NIEMAND *reads.*

P. 9. § 29. Th. 'Two angles are equal, when their sides have the same direction.'

Min. How do you define 'same direction' for different Lines?

Nie. We cannot define it.

Min. Then I cannot admit that such Lines exist. But even if I *did* admit their reality, why should the angles be equal?

Nie. Because 'the difference of direction' is the same in each case.

Min. But how would that prove the angles equal? Do you define 'angle' as the 'difference of direction' of two lines?

Nie. Not exactly. We have stated (p. 6, § 19) 'The

magnitude of the angle depends solely upon the *difference of direction* of its sides at the vertex.'

Min. But the difference of direction also possesses 'magnitude.' Is that magnitude a wholly *in*dependent entity? Or does it, in its turn, depend to some extent upon the *angle*? Seriously, all these subtleties must be very trying to a beginner. But we had better proceed to the next Theorem. I am anxious to see where, in this system, these creatures of the imagination, these sepcodal Lines, are to appear as actually existent.

Nie. We next prove (p. 9. § 30) that Lines, which have the same direction, make equal angles with all transversals.

Min. That is merely a particular case of your last Theorem.

Nie. And then that two Lines, which make equal angles with a transversal, have the same direction.

Min. Ah, *that* would bring them into existence at once! Let us hear the proof of that.

Nie. The proof is that if, through the point where the first Line is cut by the transversal, a Line be drawn having the same direction as the second, it makes equal angles with the transversal, and therefore coincides with the first Line.

Min. You assume, then, that a Line *can* be drawn through that point, having the same direction as the second Line?

Nie. Yes.

Min. That is, you assume, without proof, that different Lines can have the same direction. On the whole, then, though Mr. Pierce's system differs slightly from Mr.

Wilson's, both rest on the same vicious Axiom, that different Lines *can* exist, which possess a property called 'the same direction'—a phrase which is intelligible enough when used of two Lines which have a common point, but which, when applied to two Lines *not* known to have a common point, can neither be defined, nor constructed. We need not pursue the subject further. Have you provided any test for knowing whether two given finite Lines will meet if produced?

Nie. We have not thought it necessary.

Min. Then the only other remark I have to make on this singularly compendious treatise is that, of the 35 Theorems which Euclid gives us in his First Book, it reproduces just sixteen: the omissions being 16, 17, 25, 26 (2), 27 and 28, 29, 30, 33, 34, 35, 36, 37, 38, 39, 40, 41, 43, 47, and 48.

Nie. Most of those are in the book. For example, § 30 answers to Euc. I. 29.

Min. Only by proving that separational Lines have the same direction: which you have not done.

Nie. At any rate we have Euc. I. 47 in our § 256.

Min. Oh, no doubt! Long after going through ratios, which necessarily include incommensurables; and long after the Axiom (§ 99) 'Infinitely small quantities may be neglected'! No, no: so far as *beginners* are concerned, there is no Euc. I. 47 in *this* book!

My conclusion is that, however useful this Manual may be to an advanced student, it is *not* adapted to the wants of a beginner.

ACT II.

SCENE VI.

§ 3. WILLOCK.

'This work .... no doubt, has its faults.'
WILLOCK, *Pref.* p. 1.

Nie. I lay before you '*The Elementary Geometry of the Right Line and Circle*' by W. A. WILLOCK, D.D., formerly Fellow of Trinity College, Dublin, published in 1875.

Min. I have gone through the subject of 'direction' so minutely in reviewing Mr. Wilson's book, that I need not discuss with you any points in which your client essentially agrees with him. We may, I think, pass over the subject of the Right Line altogether?

Nie. Yes.

Min. And as to Angles and Right Angles, I see no novelty in Dr. Willock's book, except that he defines an Angle as 'the divergence of two directions,' which is virtually the same as Euclid's Definition.

Nie. That I think is all.

Min. Then we can proceed at once to the subject of Parallels. Will you kindly give me your proof of Euc. I. 32 from the beginning?

Niemand *reads.*

P. 10. Th. 1. '*Two Directives can intersect in only one point.*'

Min. By 'Directive' you mean an 'infinite Line'?

Nie. Yes.

Min. Well, I need hardly trouble you to prove it as a Theorem, being quite willing to grant it as an Axiom. What is the next Theorem?

Niemand *reads.*

P. 11. Th. 5. '*Parallel Directives cannot meet.*'

Min. We will call them 'sepcodal,' if you please. I grant it, provisionally. *If* such Lines exist, they cannot meet.

Niemand *reads.*

P. 11. Th. 7. '*Only one Line, sepcodal to a Directive, can be drawn through a point.*'

Min. Does that assert that one *can* be drawn? Or does it simply deny the possibility of drawing *two*?

Nie. The *proof* only applies to the denial: but the assertion is certainly involved in the enunciation. At all events, if not assumed here, it *is* assumed later on.

Min. Then I will at this point credit you with *one* unwarrantable Axiom, namely, that different Lines can have the same direction. The Theorem itself I grant.

NIEMAND *reads.*

P. 12. Th. 8. '*The angles of intersection of a Transversal with two sepcodal Directives are equal.*'

Min. Do you prove that by Mr. Wilson's method?

Nie. Not quite. *He* does it by transferring an angle: *we* do it by divergence of directions.

Min. I prefer *your* method. All it needs to make it complete is the proof of the reality of such Lines: but *that* is unattainable, and its absence is fatal to the whole system. Nay, more: the fact, that the reality of such Lines leads by a logical necessity to the reality of Lines which make equal angles with *any* transversal, reacts upon that unfortunate Axiom, and destroys the little hope it ever had of being granted without proof. In point of fact, in asking to have the Axiom granted, you were virtually asking to have this other reality granted as axiomatic—but all this I have already explained (p. 125).

NIEMAND *reads.*

P. 13. Th. 10. '*If a Transversal cut two Directives and make the angles of intersection with them equal, the Directives are sepcodal.*'

Min. The subject of your Proposition is indisputably real. If then you can prove this Theorem, you will thereby prove the reality of sepcodal Lines. But I fear you have assumed it already in Th. 7. There is still, however, a gleam of hope: perhaps you do not need Th. 7 in proving this?

Nie. We do not: but I fear that will not mend matters, as we assume, in the course of this Theorem, that a Line can be drawn through a given point, so as to have the same direction as a given Line.

Min. Then we need not examine it further: it must perish with the faulty Axiom on which it rests. What is your next Theorem?

Nie. It answers to Euc. I. 16, 17, and is proved by the Theorem you have just rejected.

Min. Then I must reject its *proof*, but I will grant you the Theorem itself, if you like, as we know it *can* be proved from undisputed Axioms. What comes next?

Niemand *reads.*

P. 14. Th. 13. '*If a Transversal meet two Directives, and make angles with them, the External greater than the Internal, or the sum of the two Internal angles less than two right angles, the two directives must meet.*'

Min. A proof for Euclid's Axiom? That is interesting.

Niemand *reads.*

'For, suppose they do not meet. Then, they should be sepcodal——'

Min. (*interrupting*) '*Should* be sepcodal'? Does that mean that they *are* sepcodal?

Nie. Yes, I think so.

Min. That is, you assume that separational Lines have the same direction?

Nie. We do.

Min. A fearful assumption! (*A long silence*) Well?

Nie. I am waiting to know whether you grant it.

Min. Unquestionably not! I must mark it against you as an Axiom of the most monstrous character! Mr. Wilson himself does not assume this, though he *does* assume its Contranominal, that Lines having different directions will meet (see p. 115). And what I said then I say now—unaxiomatic! But supposing it granted, how would you prove the Theorem?

NIEMAND *reads.*

'Then, they should be sepcodal; and the external angle should be equal (Th. 8) to the internal; which is contrary to the supposition.'

Min. Quite so. But Th. 8, which you quote, itself depends on the reality of sepcodal Lines. Your Theorem rests on two legs, and *both*, I fear, are rotten!

Nie. The next Theorem is equivalent to Euc. I. 32. Do you wish to hear it?

Min. It is unnecessary: it follows easily from Th. 8.

And I need not ask you what practical test you provide for the meeting of two Lines, seeing that you have Euclid's 12th Axiom itself.

Nie. Proved as a Theorem.

Min. *Attempted* to be proved as a Theorem. I will now take a hasty general survey of your client's book.

The first point calling for remark is the arrangement. You begin by dragging the unfortunate beginner straight into the most difficult part of the subject. Your first chapter positively bristles with difficulties about 'direction.' Then comes a long chapter on circles, including some very complicated figures, and a theory of tangents which depends upon moving lines and vanishing chords—all most disheartening to a beginner. What do you suppose he is likely to make of such a sentence as 'the direction of the motion of the generating point of any curve is that of the tangent to the curve at that point'? (p. 29.) Or this again, 'it is also evident that, the circle being a simple curve, there can be only *one* tangent to it at any point'? (p. 29.) What *is* 'a simple curve'?

Nie. I do not know.

Min. Then comes a chapter of Problems, and *then*—when your pupil has succeeded in mastering thirty-four pages of your book, and has become tolerably familiar with tangents and segments, with diametral lines and reëntrant angles, with 'oval forms' and 'forms semi-convex, semi-concave,'—you at last confront him with that abstruse and much dreaded Theorem, Euc. I. 4! True, he has the 'Asses' Bridge' to help him in proving it, that in its turn being proved, apparently, by properties of the circle; but, even with all these assistances, it is an arduous task!

Nie. You are hard on my client.

Min. Well, jesting apart, let me say in all seriousness that I think it would require very great ingenuity to make a worse arrangement of the subject of Geometry, for purposes of teaching, than is to be found in this little book.

I do not think it necessary to criticise the book throughout: but I will mention one or two passages which have caught my eye in glancing through it.

Here, for instance, is something about 'Directives,' which seem to be a curious kind of Loci—quite different from Right Lines, I should say.

Nie. Oh no! They are exactly the same thing!

Min. Well, I find, at p. 4, 'Directives are either divergent or parallel': and again, at p. 11, 'Parallel Directives cannot meet.' Clearly, then, Directives can never by any possibility *coincide*: but ordinary Right Lines occasionally do so, do they not?

Nie. It is a curious *lapsus pennae.*

Min. At p. 7, I observe an article headed 'The principle of double conversion,' which I will quote entire.

Reads.

'If four magnitudes, a, b, A, B, are so related, that when a is greater than b, A is greater than B; and when a is equal to b, A is equal to B: then, *conversely*, when A is greater than B, a is greater than b; and, when A is equal to B, a is equal to b.

'The truth of this principle, which extends to every kind of magnitude, is thus made evident:—If, when A is greater than B, a is not greater than b, it must be either less than or equal to b. But it cannot be less; for, if it were, A should, by the antecedent part of the proposition, be less than B, which is contrary to the supposition made. Nor can it be equal to b; for, in that case, A should be equal to B, also contrary to supposition. Since, therefore, a is neither less than nor equal to b, it remains that it must be greater than b.'

Now let a and A be variables and represent the ordinates to two curves, *mnr* and *MNR*, for the same abscissa; and let b and B be constants and represent their intercepts on the Y-axis; i.e. let $On = b$, and $ON = B$.

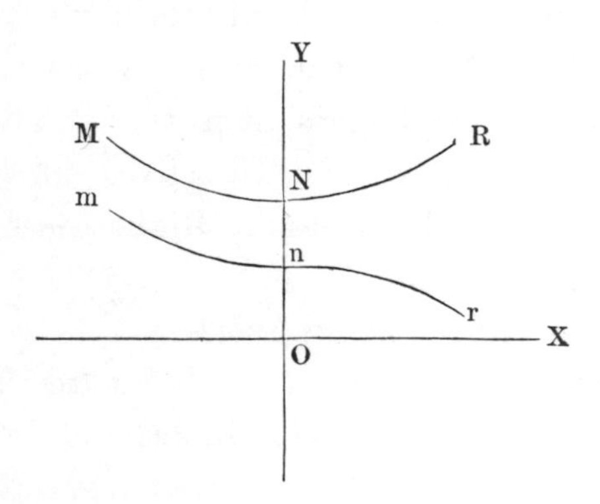

Does not this diagram fairly represent the *data* of the proposition? You see, when we take a negative abscissa, so as to make a greater than b, we are on the left-hand branch of the curve, and A is also greater than B; and again, when a is equal to b, we are crossing the Y-axis, where A is also equal to B.

Nie. It seems fair enough.

Min. But the conclusion does not follow? With a positive abscissa, A is greater than B, but a less than b.

Nie. We cannot deny it.

Min. What then do you suppose would be the effect on a simple-minded student who should wrestle with this terrible theorem, firm in the conviction that, being in a printed book, it must *somehow* be true?

Nie. (*gravely*) *Insomnia*, certainly; followed by acute *Cephalalgia*; and, in all probability, *Epistaxis*.

Min. Ah, those terrible names! Who would suppose

that a man could have all those three maladies, and survive? And yet the thing is possible!

Let me now read you a statement (at p. 112) about incommensurables:—

'When one of the magnitudes can be represented only by an interminable decimal, while the other is a finite whole number, or finite decimal, no finite common submultiple can exist; for, though a unit be selected in the last place of the whole number or finite decimal. yet the decimal represented by all the figures which follow the corresponding place in the interminable decimal, being less than that unit in that place and unknown in quantity, cannot be a common measure of the two magnitudes, and is only a remainder.'

Now can you lay your hand upon your heart and declare, on the word of an honest man, that you understand this sentence—beginning at the words 'yet the decimal'?

Nie. (*vehemently*) I cannot!

Min. Of the two reasons which are mentioned, to explain why it 'cannot be a common measure of the two magnitudes,' does the first—that it is 'less than that unit in that place'—carry conviction to your mind? And does the second—that it is 'unknown in quantity' ripen that conviction into certainty?

Nie. (*wildly*) Not in the least!

Min. Well, I will not 'slay the slain' any longer. You may consider Dr. Willock's book as rejected. And I think we may say that the whole theory of 'direction' has collapsed under our examination.

Nie. I greatly fear so.

ACT III.

SCENE I.

§ 1. THE OTHER MODERN RIVALS.

'But mice, and rats, and such small deer,
Have been Tom's food for seven long year.'

Min. I consider the question, as to whether Euclid's system and numeration should be abandoned or retained, to be now set at rest: the subject of Parallels being disposed of, no minor points of difference can possibly justify the abandonment of our old friend in favour of any Modern Rival. Still it will be worth while to examine the other writers, whose works you have brought with you, as they may furnish some valuable suggestions for the improvement of Euclid's Manual.

Nie. The other writers are CHAUVENET, LOOMIS, MORELL, REYNOLDS, and WRIGHT.

Min. There are a few matters, as to which we may consider them all at once. How do they define a straight Line?

Nie. All but Mr. Reynolds define it as the shortest distance between two points, or more accurately, to use the words of Mr. Chauvenet, 'a Line of which every portion is the shortest Line between the points limiting that portion.

Min. We discussed that Definition in M. Legendre's book. How does Mr. Reynolds define it?

Nie. Not at all.

Min. Very cautious. What of angles?

Nie. Some of them allow larger limits than Euclid does. Mr. Wright talks about 'angles of continuation' and 'angles of rotation.'

Min. Good for Trigonometry: not so suitable to early Geometry. How do they define Parallels?

Nie. As in Euclid, all of them.

Min. And which Proposition of Tab. II. do they assume?

Nie. Playfair's, or else its equivalent, 'only one Line can be drawn, parallel to a given Line, through a given point outside it.'

Min. Now let us take them one by one.

ACT III.

SCENE I.

§ 2. CHAUVENET.

'Where Washington hath left
His awful memory
A light for after times!'

Nie. I lay before you '*A Treatise on Elementary Geometry*,' by W. CHAUVENET, LL.D., Professor of Mathematics and Astronomy in Washington University, published in 1876.

Min. I read in the Preface (p. 4) 'I have endeavoured to set forth the elements with all the rigour and completeness demanded by the present state of the general science, *without seriously departing from the established order of the Propositions*.' So there would be little difficulty, I fancy, in introducing into Euclid's own Manual all the improvements which Mr. Chauvenet can suggest.

P. 14. Pr. I, and p. 18. Pr. V, taken together, tell us that only one perpendicular can be drawn to a Line from a point. And various additions, about obliques, are made in subsequent Propositions. All these may well be embodied in a new Proposition, which we might interpolate as Euc. I. 12. B.

P. 26. Pr. XV, asserts the equidistance of Parallels. This might be interpolated as Euc. I. 34. B.

Another new Theorem, that angles whose sides are parallel, each to each, are equal (which I observe is a great favourite with the Modern Rivals), seems to me a rather clumsy and uninteresting extension of Euc. I. 29.

I see several Propositions which might well be inserted as *exercises* on Euclid (*e.g.* Pr. XXXIX, 'Every point in the bisector of an angle is equally distant from the sides'), but which are hardly of sufficient importance to be included as Propositions: and others (*e.g.* Pr. XL, 'The bisectors of the three angles of a Triangle meet in the same point) which seem to belong more properly to Euc. III or IV. I have no other remarks to make on this book, which seems well and clearly written.

ACT III.

SCENE I.

§ 3. LOOMIS.

'Like—but oh! how different!'

Nie. I lay before you '*Elements of Geometry*,' by ELIAS LOOMIS, LL.D., Professor of Natural Philosophy and Astronomy in Yale College, a revised edition, 1876.

Min. I read in the Preface (p. 10) 'The present volume follows substantially the order of Blanchet's Legendre, while the form of the demonstrations is modeled after the more logical method of Euclid.' He has not, however, adopted the method of infinite series, which constitutes the crucial distinction between that writer and Euclid.

The Propositions are pretty nearly in Euclid's order: with a few changes in order and numeration, the book would be a modernised Euclid, the only important differences being the adoption of Playfair's Axiom and the omission of the diagonals in Euc. II. I have no hostile criticisms to offer. Our American cousins set us an excellent example in the art of brief, and yet lucid, mathematical writing.

ACT III.

SCENE I.

§ 4. MORELL.

'Quis custodiet ipsos custodes?
Quis inspiciet ipsos inspectores?'

Nie. I lay before you '*Euclid Simplified, compiled from the most important French works, approved by the University of Paris and the Minister of Public Instruction,*' by Mr. J. R. MORELL, formerly H. M. Inspector of Schools, published in 1875.

Min. What have you about Lines, to begin with?

Nie. Here is a Definition. 'The place where two surfaces meet is called a Line.'

Min. Really! Let us take two touching spheres, for instance?

Nie. Ahem! We abandon the Definition.

Min. Perhaps we shall be more fortunate with the Definition of a *straight* Line.

Nie. It is 'an indefinite Line, which is the shortest between any two of its points.'

Min. An '*indefinite*' Line! What in the world do you mean? Is a curved Liue more definite than a straight Line?

Nie. I don't know.

Min. Nor I. The rest of the sentence is slightly elliptical. Of course you mean 'the shortest which can be drawn'?

Nie. (*eagerly*) Yes, yes!

Min. Well, we have discussed that matter already. Go on.

Nie. Next we have an Axiom, 'that from one point to another only one straight Line can be drawn, and that if two portions of a straight Line coincide, these Lines coincide throughout their whole extent.'

Min. You bewilder me. How can one portion of a straight Line coincide with another?

Nie. (*after a pause*) It can't, of course, *in situ*: but why not take up one portion and lay it on another?

Min. By all means, if you like. Let us take a certain straight Line, cut out an inch of it, and lay it along another inch of the Line. What follows?

Nie. Then 'these Lines coincide throughout their whole extent.'

Min. Do they indeed? And pray who *are* 'these Lines'? The two inches?

Nie. (*gloomily*) I suppose so.

Min. Then the Axiom is simple tautology.

Nie. Well then, we mean the whole straight Line and—and—

Min. And what else? You can't talk of 'one straight Line' as 'these Lines,' you know.

Nie. We abandon the Axiom.

Min. Better luck next time! Try another Definition.

Nie. 'A broken Line is a Line composed of straight Lines.'

Min. But a *straight* Line also is 'a Line composed of straight Lines,' isn't it?

Nie. Well, we abandon the Definition.

Min. This is quite a new process in our navigation. Instead of heaving the lead, we seem to be throwing overboard the whole of our cargo! Let us hear something about Angles.

Nie. 'The figure formed by two Lines that intersect is called an Angle.'

Min. What do you mean by 'figure'? Do you define it anywhere?

Nie. Yes. 'The name of figure is given to volumes, surfaces, and lines.'

Min. Under which category do you put 'Angle'?

Nie. I don't know.

Min. Anything new about the Definition, or equality, of right angles?

Nie. No, except that we *prove* that all right angles are equal.

Min. That we have discussed already (see p. 57). Let us go on to Pairs of Lines, and your proof of Euc. I. 29, 32.

NIEMAND *reads.*

'Th. 19. Two Lines perpendicular to the same Line are parallel.'

Min. Do you mean 'separational'?

Nie. Yes.

Min. Have you defined 'parallel' anywhere?

Nie. (*after a search*) I can't find it.

Min. A careless omission. Moreover, your assertion isn't always true. Suppose your two Lines were drawn from the same point?

Nie. We beg to correct the sentence. 'Two *different* Lines.'

Min. Very well. Then you assert Table I. 6. (See p. 29.) I grant it.

NIEMAND *reads.*

'Th. 20. Through a point situated outside a straight Line a Parallel, and only one, can be drawn to that Line.'

Min. '*A* Parallel,' I grant at once: it is Table I. 9. But '*only one*'! That takes us into Table II. What axiom do you assume?

Nie. 'It may be admitted that only one Parallel can be drawn to it.'

Min. That is Table II. 15 (*b*)—a contranominal of Playfair's Axiom. We need not pursue the subject: all is easy after that. Now hand *me* the book, if you please: I wish to make a general survey of style, &c.

At p. 4 I read:—'Two Theorems are reciprocal when the hypothesis and the conclusion of one are the conclusion and the hypothesis of the other.' (They are usually called 'converse'—the *technical*, not the *logical*, converse, as was mentioned some time ago (p. 47); but let that pass.)

'Thus the Theorem—*if two angles are right angles, they are equal*—has for its reciprocal—*if two angles are equal, they are right angles.*'

(This, by the way, is a capital instance of the distinction between 'technical' and 'logical.' Here the *technical* converse is wild nonsense, while the *logical* converse is of course as true as the Theorem itself: it is '*some cases of two angles being equal are cases of their being right.*')

'All Propositions are direct, reciprocal, or contrary—all so closely connected that either of the two latter' (I presume he means 'the latter two') 'is a consequence of the other two.'

A 'consequence'! Can he mean a *logical* consequence? Would he let us make a syllogism of the three, using the 'direct' and 'reciprocal' (for instance) as premisses, and the 'contrary' as the conclusion?

However, let us first see what he means by a 'contrary' Proposition.

'It is a direct Proposition to prove that all points in a circle enjoy a certain property, *e.g.* the same distance from the centre.'

(This notion of sentient points, by the way, is very charming. I like to think of all the points in a circle really feeling a placid satisfaction in the thought that they are equidistant from the centre! They are infinite in number, and so can well afford to despise the arrogance of a point within, and to ignore the envious murmurs of a point without!)

'The contrary Proposition shows that all points taken outside or inside the figure do not enjoy this property.'

So then this is his trio :—

1. Direct. 'All *X* are *Y*.'
2. Reciprocal. 'All *Y* are *X*.'
3. Contrary. 'All not-*X* are not-*Y*.'

Here of course No. 2 and No. 3, being Contranominals, are logically deducible from each other, No. 1 having no logical connection with either of them.

And yet he calls the three 'so closely connected that either of the two latter is a consequence of the other two'! Shade of Aldrich! Have we come to this? You say nothing, mein Herr?

Nie. I say that, if you grant what you call the 'premisses,' you cannot deny the conclusion.

Min. True. It reminds me of an answer given some years ago in the Schools at Oxford, when the Examiner asked for an example of a syllogism. After much patient thought, the candidate handed in

'All men are dogs;
All dogs are men:
Therefore, All men are dogs.'

This certainly has the *form* of a syllogism. Also it avoids, with marked success, the dangerous fallacy of 'four terms.' And it has the great merit of Mr. Morell's syllogism, that, if you grant the premisses, you cannot deny the conclusion. Nevertheless I feel bound to add that it was *not* commended by the Examiner.

Nie. I can well believe it.

Min. I proceed. 'The direct and the reciprocal proofs are generally the simpler, and do not require a fresh construction.' Why 'fresh'? The 'direct' comes *first*, apparently; so that, if it requires a construction at all, it *must* be a 'fresh' one.

Nie. Be not hypercritical.

Min. Well, it *is* rather 'small deer,' I confess: let us change the subject.

Here is a pretty proof in Th. 4.

'Then $m + o = m + x$.
But $m = m$.
Therefore $o = x$.'

Isn't that 'but $m = m$' a delightfully cautious parenthesis? Your client seems to be nearly as much at home in Algebra as in Logic, which is saying a great deal!

At p. 9, I read 'The base of an isosceles Triangle is the unequal side.'

'*The unequal side*'! Is an equilateral Triangle isosceles, or is it not? Answer, mein Herr!

Nie. Proceed.

Min. At p. 17, I read 'From one and the same point three equal straight Lines cannot be drawn to another straight Line; for if that were the case, *there would be on the same side of a perpendicular two equal obliques,* which is impossible.'

Kindly prove the italicised assertion on this diagram, in which I assume *FD, FC, FE,* to be equal Lines, and

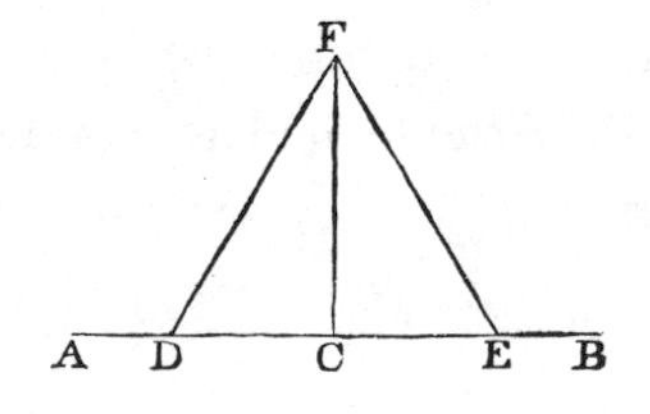

have made the *middle* one of the three a perpendicular to the 'other straight Line.'

Nie. (*furiously*) I will not!

Min. Look at p. 36. 'A circumference is generally described in language by one of its radii.' Let us hope that the language is complimentary—at least if the circumference is within hearing! Can't you imagine the radius gracefully rising to his feet, rubbing his lips with his table-napkin? 'Gentlemen! The toast I have the honour to propose is &c. &c. Gentlemen, I give you *the Circumference*!' And then the chorus of excited Lines, 'For he's a jolly good felloe!'

Nie. (*rapturously*) Ha, ha! (*checking himself*) You are insulting my client.

Min. Only filling in his suggestive outlines. Try p. 48. 'Th. 13. If two circumferences are interior,' &c. Can your imagination, or mine, grasp the idea of two circumferences, each of them inside the other? No! *We* are mere prosaic mortals: it is beyond us!

In p. 49 I see some strange remarks about ratios. First look at Def. 44. 'When a magnitude is contained an exact number of times in two magnitudes of its kind, it is said to be their common measure.' (The wording is awkward, and suggests the idea of their having only *one* 'common measure'; but let that pass.) 'The ratio of two magnitudes of the same kind is the number which would express the measure of the first, if the second were taken as unity.'

'*The measure of the first*'! Do you understand that? Is it a 'measure' such as you have just defined? or some other kind?

Nie. Some other kind, I *think.* But there is a slight obscurity somewhere.

Min. Perhaps this next enunciation will clear it up. 'If two magnitudes of the same kind, *A* and *B*, are mutually commensurable' (by the way, 'mutually' is tautology), 'their ratio is a whole or fractional number, which is obtained by dividing the two numbers one by the other, and which expresses how many times these magnitudes contain their common measure *M.*' Do you understand *that*?

Nie. Well, no!

Min. Let us take an instance—£3 and 10*s.* A shilling is *a* common measure of these two sums: will you accept it as '*their* common measure'?

Nie. We will do it, provisionally.

Min. Now the number, 'obtained by dividing the two numbers' (I presume you mean 'the two magnitudes') 'one by the other,' is '6,' is it not?

Nie. It would seem so.

Min. Well, does this number 'express how many times these magnitudes contain their common measure,' viz. a shilling?

Nie. Hardly.

Min. Did you ever meet with any *one* number that could 'express' *two* distinct facts?

Nie. We would rather change the subject.

Min. Very well, though there is plenty more about it, and the obscurity deepens as you go on. We will 'vary the verse' with a little bit of classical criticism. Look at p. 81. 'Homologous, from the Greek ὁμοῖος, like or similar, λόγος, word or reason.' Do you think this school-inspector ever heard of the great Church controversy, where all turned on the difference between ὅμος and ὁμοῖος?

Nie. (*uneasily*) I think not. But this is not a *mathematical* slip, you know.

Min. You are right. *Revenons à nos moutons.* Turn to p. 145, art. 65. 'To measure areas, it is usual to take a square as unity.' To me, who have always been accustomed to regard 'a square' as a concrete magnitude and 'unity' as a pure number, the assertion comes rather as a shock. But I acquit the author of any intentional roughness. Nothing could surpass the delicacy of the next few words:—'It has been already stated that surfaces are measured indirectly'! Lines, of course, may be measured anyhow: *they* have no sensibilities to wound: but there is an open-handedness—a breadth of feeling—about a surface, which tells of noble birth—'every (square) inch a King!'—and so we measure it with averted eyes, and whisper its area with bated breath!

Nie. Return to other muttons.

Min. Well, take p. 156. Here is a 'scholium' on a theorem about the area of a sector of a circle. The 'scholium' begins thus:—'If a is the number of degrees in the arc of a sector, we shall have to find the length of this arc ——.' I pause to ask 'If β were the number, should we have to find it *then*?'

Nie. (*solemnly*) We should!

Min. 'For the two Lines which are multiplied in all rules for the measuring of areas must be referred to the same linear unity.' *That*, I take it, is fairly obscure: but it is luminous when compared with the note which follows it. 'If the linear unit and angular unit are left arbitrary, any angle has for measure the ratio of the numbers of linear units contained in the arcs which the angle in question and the irregular unit intercept in any circumference described from their summit as common centres.' Is not that a useful note? 'The irregular unit'! Linear, or angular, I wonder? And then 'common centres'! How many centres does a circumference usually require? I will only trouble you with one more extract, as a *bonne bouche* to wind up with.

'Th. 9. (P. 126.) *Every convex closed Line* ABCD *enveloped by any other closed Line* PQRST *is less than it.*

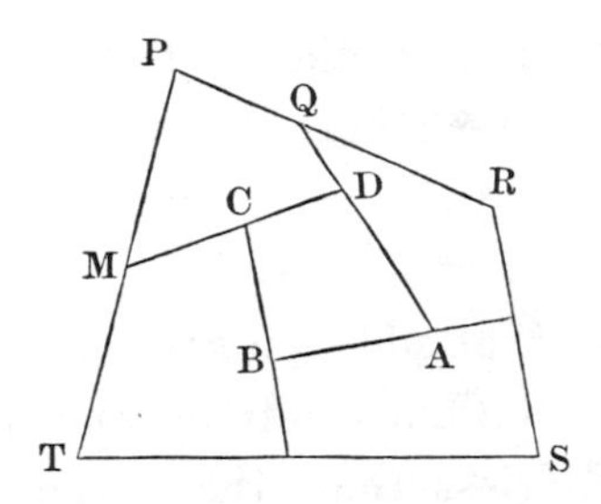

'All the infinite Lines *ABCD*, *PQRST*, &c.'—— by the way, these are curious instances of 'infinite Lines'?

Nie. (*hastily*) We mean 'infinite' in *number*, not in *length*.

Min. Well, you express yourself oddly, at any rate '—— which enclose the plane surface *ABCD*, cannot be equal. For drawing the straight Line *MD*, which does not cut *ABCD*, *MD* will be less than *MPQD*; and adding to both members the part *MTSRQD*, the result will be *MDQRSTM* less than *MPQRSTM*.' Is that result proved?

Nie. No.

Min. Is it true?

Nie. Not necessarily so.

Min. Perhaps it is a *lapsus pennæ*. Try to amend it.

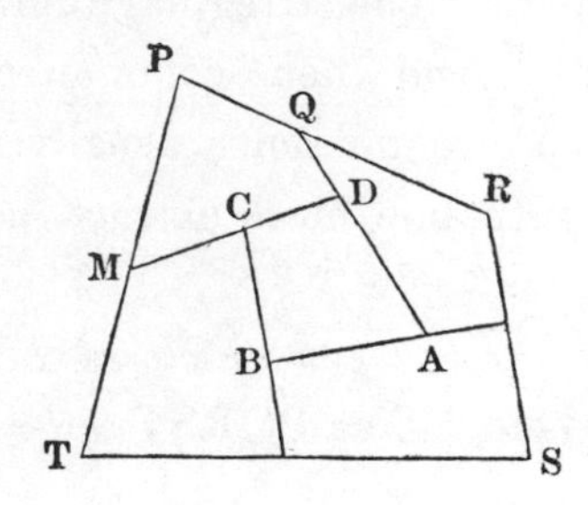

Nie. If we add to *MD* the part *MTSRQD*, we do get *MDQRSTM*, it is true: but, if we add it to *MPQD*, we get *QD* twice over; that is, we get *MPQRSTM* together with twice *QD*.

Min. How does that addition suit the rest of the proof?

Nie. It ruins it: all depends on our proving the perimeter *MDQRSTM* less than the perimeter *MPQRSTM*, which this method has failed to do—as of course all methods must, the thing not being capable of proof.

Min. Then the whole proof breaks down entirely?

Nie. We cannot deny it.

Min. Let us turn to the next author.

ACT III.

SCENE I.

§ 5. REYNOLDS.

'Though this be madness, yet there's method in 't.'

Nie. I lay before you '*Modern Methods in Elementary Geometry*,' by E. M. REYNOLDS, M.A., Mathematical Master in Clifton College, Modern Side; published in 1868.

Min. The first remark I have to make on it is, that the Definitions and Axioms are scattered through the book, instead of being placed together at the beginning, and that there is no index to them, so that the reader only comes on them by chance: it is quite impossible to refer to them.

Nie. I cannot defend the innovation.

Min. In Th. 1 (p. 3), I read 'the angles *CDA*, *CDB* are together equal to two right angles. *For they fill exactly the same space*.' Do you mean finite or infinite space? If 'finite,' we increase the angle by lengthening its sides: if 'infinite,' the idea is unsuited for elementary teaching. You had better abandon the idea of an angle 'filling space,' which is no improvement on Euclid's method.

P. 61. Th. II (of Book III) it is stated that Parallelograms, on equal bases and between the same Parallels, 'may always be placed so that their equal bases coincide,' and it is clearly assumed that they will still be 'between the same Parallels.' And again, in p. 63, the altitude of a Parallelogram is defined as '*the* perpendicular distance of the opposite side from the base,' clearly assuming that there is only *one* such distance. In both these passages the Theorem is assumed 'Parallels are equidistant from each other,' of which no proof has been given, though of course it might have been easily deduced from Th. XVI (p. 19).

The Theorems in Euc. II are here proved algebraically, which I hold to be emphatically a change for the worse, chiefly because it brings in the difficult subject of incommensurable magnitudes, which should certainly be avoided in a book meant for beginners.

I have little else to remark on in this book. Several of the new Theorems in it seem to me to be premature, e.g. Th. XIX, &c. on 'Loci': but the sins of *omission* are more serious. He actually leaves out Euc. I. 7, 17, 21 (2nd part), 24, 25, 26 (2nd part), 48, and II. 1, 2, 3, 8, 9, 10, 12, 13. Moreover he separates Problems and Theorems, which I hold to be a mistake. I will not trouble you with any further remarks.

ACT III.

Scene I.

§ 6. Wright.

'Defects of execution unquestionably exist.'
Wright, *Pref.* p. 10

Nie. I lay before you '*The Elements of Plane Geometry*,' by R. P. Wright, Teacher of Mathematics in University College School, London; the second edition, 1871.

Min. Some of the changes in Euclid's method, made in this book, are defended in the Preface.

First, he claims credit for having more Axioms than Euclid, whom he blames for having demonstrated 'much that is obvious.' I need hardly pause to remind you that 'obviousness' is not an invariable property: to a *perfect* intellect the whole of Euclid, to the end of Book XII, would be 'obvious' as soon as the Definitions had been mastered: but Geometricians must write for *imperfect* intellects, and it cannot be settled on general principles where Axioms should end and Theorems begin. Let us look at a few of these new Axioms. In p. viii of the Preface, I read 'with the conception of straightness in a Line we naturally associate that of the utmost possible shortness of path between any two of its points; allow this to be assumed, &c.' This I consider a most objectionable Axiom, obliging us, as it does, to contemplate the lengths of *curved* lines. This matter I have already discussed with M. Legendre (p. 56).

Secondly, for the host of new Axioms with which we are threatened in the Preface, I have searched the book in vain: possibly I have overlooked some, as he never uses the heading 'Axiom,' but really I can only find *one* new one, at p. 5. 'Every angle has one, and only one bisector,' which is hardly worth stating. Perhaps the writer means that his proofs are not so full as those in Euclid, but take more for granted. I do not think this any improvement in a book meant for beginners.

Another change, claimed in the Preface as an improvement, is the more constant use of superposition. I have considered that point already (p. 47) and have come to the conclusion that Euclid's method of constructing a new figure has all the advantages, without the obscurity, of the method of superposition.

I see little to remark on in the general style of the book. At p. 21 I read 'the straight Line *AI* satisfies the four following conditions: it passes through the vertex *A*, through the middle point *I* of the base, is a perpendicular on that base, and is the bisector of the vertical angle. Now, two of these four conditions suffice to determine the straight Line *AI*, . . . Hence a straight Line fulfilling any two of these four conditions necessarily fulfils the other two.' All this is strangely inaccurate: the fourth condition is sufficient by itself to determine the line *AI*.

At p. 40 I notice the startling announcement that 'the simplest of all Polygons is the *Triangle*'! This is surely a new use for 'many'? I wonder if the writer is prepared to accept the statement that '*many* people have swum across the Bosphorus' on the strength of Byron's

'As once (a feat on which ourselves we prided)
Leander, Mr. Ekenhead, and I did.'

As a specimen of the wordy and unscientific style of the writer, take the following:—

'*From any point O, one, and only one, perpendicular can be drawn to a given straight Line AB.*

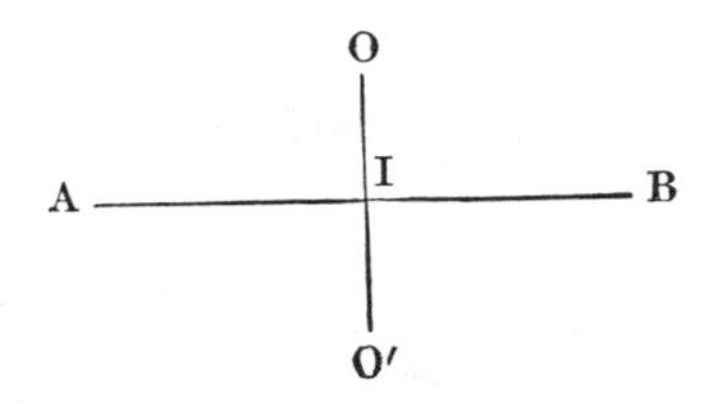

'Let O' be the point on which O would fall if, the paper being folded along AB, the upper portion of the figure were turned down upon the lower portion. If from the points O, O' straight Lines be drawn to any point whatever I on the line AB, the adjacent angles OIB, $O'IB$ will be equal; for folding the paper again along AB and turning the upper portion down upon the lower, O falls on O', I remains fixed, and the angle OIB exactly coincides with $O'IB$. Now in order that the Line OI may be perpendicular to AB, or, in other words, that OIB may be a right angle, the sum of the two adjacent angles OIB, $O'IB$ must be equal to two right angles, and consequently their sides IO, IO' in the same straight Line. But since we can always draw one, and only one, straight Line between two points O and O', it follows that from a point O we can always draw one, and only one, perpendicular to the line AB.'

Do you think you could make a more awkward or more obscure proof of this almost axiomatic Theorem?

Nie. (*cautiously*) I would not undertake it.

Min. All that about folding and re-folding the paper is more like a child's book of puzzles than a scientific treatise. I should be very sorry to be the school-boy who is expected to *learn* this precious demonstration! In such a case, I could not better express my feelings than by quoting three words of this very Theorem :—'*I remains fixed*'!

In conclusion, I may say as to all five of these authors, that they do not seem to me to contain any desirable novelty which could not easily be introduced into an amended edition of Euclid.

Nie. It is a position I cannot dispute.

ACT III.

SCENE II.

§ 1. SYLLABUS OF THE ASSOCIATION FOR THE IMPROVEMENT OF GEOMETRICAL TEACHING. 1878.

'Nos numerus sumus.'

Nie. The last book to be examined is Mr. Wilson's new Manual, founded on the Syllabus of the Geometrical Association.

Min. We had better begin by examining the Syllabus itself. I own that I could have wished to do this in the presence of some member of the Committee, who might have supplied a few details for what is at present little more than a skeleton, but that I fear is out of the question.

Nie. Nay, you shall not have far to seek. *I* am a member of the Committee.

Min. (*astonished*) You! A German professor! No such member is included in the final list of the Committee, which a friend showed me the other day.

Nie. The final list, was it? Well, ask your friend whether, since the drawing up of that list, any addition has been made: he will say 'Nobody has been added.'

Min. Quite so.

Nie. You do not understand. *Nobody—Niemand*—see you not?

Min. What? You mean—

Nie. (*solemnly*) I do, my friend. *I* have been added to it!

Min. (*bowing*) The Committee are highly honoured, I am sure.

Nie. So they ought to be, considering that I am a more distinguished mathematician than Newton himself, and that *my* Manual is better known than Euclid's! Excuse my self-glorification, but any moralist will tell you that I—I alone among men—*ought* to praise myself.

Min. (*thoughtfully*) True, true. But all this is word-juggling—a most misleading analogy. However, as you now appear in a new character, you must at least have a new name!

Nie. (*proudly*) Call me *Nostradamus!*

[*Even as he utters the mystic name, the air grows dense around him, and gradually crystallizes into living forms. Enter a phantasmic procession, grouped about a banner, on which is emblazoned in letters of gold the title* 'ASSOCIATION FOR THE IMPROVEMENT OF THINGS IN GENERAL.' *Foremost in the line marches* NERO, *carrying his unfinished 'Scheme for lighting and warming Rome'; while among the crowd which follow him may be noticed*—GUY FAWKES, *President of the 'Association for raising the position of Members of Parlia-*

ment'—THE MARCHIONESS DE BRINVILLIERS, *Inventress of the 'Application of Alteratives to the Digestive Faculty'—and* THE REV. F. GUSTRELL (*the being who cut down Shakspeare's mulberry-tree*), *leader of the 'Association for the Refinement of Literary Taste.' Afterwards enter, on the other side, Sir Isaac Newton's little dog* '*DIAMOND*,' *carrying in his mouth a half-burnt roll of manuscript. He pointedly avoids the procession and the banner, and marches past alone, serene in the consciousness that he, single-pawed, conceived and carried out his great 'Scheme for throwing fresh light on Mathematical Research,' without the aid of any Association whatever.*]

Min. Nostra, the plural of *nostrum*, 'a quack remedy'; and *damus*, 'we give.' It is a suggestive name.

Nos. And, trust me, it is a suggestive book that I now lay before you. '*Syllabus*—'.

Min. (*interrupting*) You mean '*a* Syllabus', or '*the* Syllabus'?

Nos. No, no! In this railroad-age, we have no time for superfluous words! '*Syllabus of Plane Geometry, prepared by the Association for the Improvement of Geometrical Teaching.*' Fourth Edition, 1877.

Min. How do you define a Right Line?

NOSTRADAMUS *reads.*

P. 7. Def. 5. 'A straight line is such that any part will, however placed, lie wholly on any other part, if its extremities are made to fall on that other part.'

Min. That looks more like a *property* of a Right Line than its *essence*. Euclid makes an Axiom of that property. Of course you omit his Axiom?

Nos. No. We have the Axiom (p. 10, Ax. 2) 'Two straight lines that have two points in common lie wholly in the same straight line.'

Min. Well! That is certainly the strangest Axiom I ever heard of! The idea of asserting, as an Axiom, that Right Lines answer to their Definition!

Nos. (*bashfully*) Well, you see there were several of us at work drawing up this Syllabus: and we 've got it a little mixed: we don't quite know which are Definitions and which are Axioms.

Min. So it appears: not that it matters much: the practical test is the only thing of importance. Do you adopt Euc. I. 14?

Nos. Yes.

Min. Then we may go on to the next subject. Be good enough to define 'Angle.'

NOSTRADAMUS *reads.*

P. 8. Def. 11. 'When two straight lines are drawn from the same point, they are said to contain, or to make with each other, a *plane angle.*'

Min. Humph! You are very particular about drawing them *from* a point. Suppose they were drawn *to* the same point, what would they make then?

Nos. An angle, undoubtedly.

Min. Then why omit that case? However, it matters little. You say '*a* plane angle,' I observe. You limit an angle, then, to a magnitude less than the sum of two right angles.

Nos. No, I can't say we do. A little further down we

assert that '*two* angles are formed by two straight lines drawn from a point.'

Min. Why, these are like Falstaff's 'rogues in buckram suits'! Are there more coming?

Nos. No, we do not go beyond the sum of four right angles. These two we call *conjugate* angles. 'The greater of the two is called the *major conjugate*, and the smaller the *minor conjugate*, angle.'

Min. These Definitions are wondrous! This is the first time I ever heard 'major' and 'minor' defined. One feels inclined to say, like that Judge in the story, when a certain barrister, talking against time, insisted on quoting authorities for the most elementary principles of law, 'Really, brother, there are *some* things the Court may be assumed to know!' Any more definitions?

Nos. We define 'a straight angle.'

Min. That I have discussed already (see p. 102).

Nos. But *this*, I think, is new :—

Reads.

P. 9. Def. 12. 'When three straight Lines are drawn from a point, if one of them be regarded as lying between the other two, the angles which this one (the mean) makes with the other two (the extremes) are said to be *adjacent* angles.'

Min. That is new indeed. Let us try a figure :—

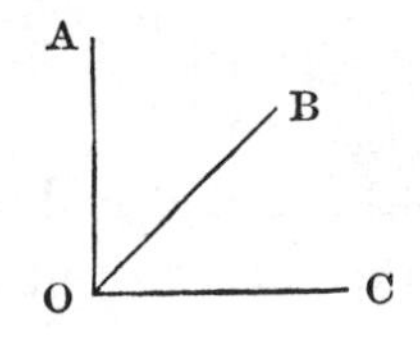

Now let us regard *OA* 'as lying between the other two.' Which are '*the* angles which it makes with the other two'? For this line *OA* (which you rightly call 'the mean'—lying is always mean) makes, be pleased to observe, *four* angles altogether—two with *OB*, and two with *OC*.

Nos. I cannot answer your question. You confuse me.

Min. I need not have troubled you. I see that I can obtain an answer from the Syllabus itself. It says (at the end of Def. 11) 'when *the angle contained by two lines* is spoken of without qualification, the *minor conjugate* angle is to be understood.' Here we have a case in point, as *these* angles are spoken of 'without qualification.' So that the angles alluded to are both of them 'minor conjugate' angles, and lie on the same side of *OA*. And these we are told to call 'adjacent' angles!

How do you define a Right Angle?

Nos. As in Euclid.

Min. Let me hear it, if you please. You know Euclid has no major or minor conjugate angles.

NOSTRADAMUS *reads*.

P. 9. Def. 14. 'When one straight line stands upon another straight line and makes the adjacent angles equal, each of the angles is called a *right angle*.'

Min. Allow me to present you with a figure, as I see the Syllabus does not supply one.

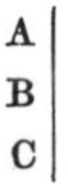

A
B
C

Here *AB* 'stands upon' *BC* and makes the adjacent angles equal. How do you like these 'right angles'?

Nos. Not at all.

Min. These same 'conjugate angles' will get you into many difficulties.

Have you Euclid's Axiom 'all right angles are equal'?

Nos. Yes; only *we* propose to prove it as a Theorem.

Min. I have no objection to that: nor do I think that your treatment of angles, as a whole, is actually *illogical.* What I chiefly object to is the general 'slipshoddity' (if I may coin a word) of the language of your Syllabus.

Does your proof of Euc. I. 32 differ from his?

Nos. No, except that we propose Playfair's Axiom, 'two straight Lines that intersect one another cannot both be parallel to the same straight Line,' as a substitute for Euc. Ax. 12.

Min. Is this your only test for the meeting of two Lines, or do you provide any other?

Nos. This is the only one.

Min. But there are cases where this is of no use. For instance, if you wish to make a Triangle, having, as *data*, a side and the two adjacent angles. Have you such a Problem?

Nos. Yes, it is Pr. 10, at p. 19.

Min. And how do you prove that the Lines will meet?

Nos. (*smiling*) We *don't* prove it: that is the reader's business: we only provide enunciations.

Min. You are like the *gourmand* who would eat so many oysters at supper that at last his friend could not help saying 'They are sure to disagree with you in the night.' 'That is *their* affair,' the other gaily replied. '*I* shall be asleep!'

Your Syllabus has the same hiatus as the other writers who have rejected Euclid's 12th Axiom. If you will not have it as an Axiom, you ought to prove it as a Theorem. Your treatise is incomplete without it.

13—15
4, 5
26 *a*
6
16
18—24
8
25
26 *β*
17

The Theorems contained in the first 26 Propositions of Euclid are thus rearranged in the Syllabus. The only advantage that I can see in the new arrangement is that it places first the three which relate to Lines, thus getting all those which relate to Triangles into a consecutive series. All the other changes seem to be for the worse, and specially the separation of Theorems from their converses, *e.g.* Props. 5, 6, and 24, 25.

The third part of Prop. 29 is put after Prop. 32: and Props. 33, 34 are transposed. I can see no reason for either change.

Prop. 47 is put next before Prop. 12 in Book II. This would be a good arrangement (if it were ever proved to be worth while to abandon Euclid's order), as the Theorems are so similar; and the placing Prop. 48 next after II. 13 is a necessary result.

In Book II, Props. 9, 10 are placed after Props. 12, 13. I see no reason for it.

It does not appear to me that the new arrangements, for the sake of which it is proposed to abandon the numeration of Euclid, have anything worth mentioning to offer as an advantage.

I will now go through a few pages of 'this many-headed monster,' and make some general remarks on its *style*.

P. 4. 'A *Theorem* is the formal statement of a Proposition that may be demonstrated from known Propositions. These known Propositions may themselves be Theorems or Axioms.'

This is a truly delightful jumble. Clearly, 'a Proposition that may be demonstrated from known Propositions' is itself a Theorem. Hence a Theorem is 'the formal statement' *of* a Theorem. The question now arises—of itself, or of some other Theorem? That a Theorem should be 'the formal statement' of *itself*, has a comfortable domestic sound, something like 'every man his own washerwoman,' but at the same time it involves a fearful metaphysical subtlety. That one Theorem should be 'the formal statement' of *another* Theorem, is, I think, degrading to the former, unless the second will consent to act on the 'claw me, claw thee' principle, and to be 'the formal statement' of the first.

Nos. You bewilder me.

Min. Perhaps, however, it is intended that the teacher who uses this Manual should, on reaching the words 'a Proposition that may be demonstrated,' recognise the fact that this is itself 'a Theorem,' and at once go back to the beginning of the sentence. He will thus obtain a Definition closely resembling a Continued Fraction, and may go on repeating, as long as his breath holds out, or until his pupil declares himself satisfied, 'a *Theorem* is the formal statement of the formal statement of the formal statement of the ——'

Nos. (*wildly*) Say no more! My brain reels!

Min. I spare you. Let us go on to p. 5, where I find the following:—

'*Rule of Conversion.* If of the hypotheses of a group of demonstrated Theorems it can be said that one must be true, and of the conclusions that no two can be true at the same time, then the converse of every Theorem of the group will necessarily be true.'

Let us take an instance:—

If $5 > 4$, then $5 > 3$.
If $5 < 2$, then $5 < 3$.

Those will do for 'demonstrated Theorems,' I suppose?

Nos. I suppose so.

Min. And the 'hypothesis' of the first 'must be true,' simply because it *is* true.

Nos. It would seem so.

Min. And it is quite clear that 'of the conclusions no two can be true at the same time,' for they contradict each other.

Nos. Clearly.

Min. Then it ought to follow that 'the converse of every Theorem of the group will necessarily be true.' Take the converse of the second, i. e.

If $5 < 3$, then $5 < 2$.

Is this 'necessarily true'? Is every thing which is less than 3 necessarily less than 2?

Nos. Certainly not. I think you have misinterpreted the phrase 'it can be said that one must be true,' when used of the hypotheses. It does not mean 'it can be said, from a knowledge of the subject-matter of some *one* hypothesis, that it *is*, and therefore must be, true,' but 'it can be said, from a knowledge of the mutual logical relation of *all* the hypotheses, as a question of *form* alone, and without any knowledge of their subject-matter, that one must be true, though we do not know which it is.'

Min. Your power of uttering long sentences is one that does equal honour to your head and your—lungs. And most sincerely do I pity the unfortunate learner who has to make out all that for himself! Let us proceed.

P. 9. Def. 13. 'The *bisector* of an angle is the straight Line that divides it into two equal angles.'

This assumes that 'an angle has one and only one bisector,' which appears as Ax. 4, at the foot of p. 10.

P. 10. Def. 21. 'The opposite angles made by two straight Lines that intersect &c.'

This seems to imply that 'two Lines that intersect' always *do* make 'opposite angles.'

Nos. Surely they do?

Min. By no means. Look at p. 12, Def. 32, where, in speaking of a Triangle, you say 'the intersection of the other two sides is called the vertex.'

Nos. A slip, I confess.

Min. One of many.

P. 12, Def. 31. 'All other Triangles are called acute-angled Triangles.' What? If a Triangle had two right angles, for instance?

Nos. But there *is* no such Triangle.

Min. *That* is a point you do not prove till we come to Th. 18, Cor. 1, two pages further on. The same remark applies to your Def. 33, in the same page. 'The side . . .

which is opposite to *the* right angle,' where you clearly assume that it cannot have more than one.

P. 12, Def. 32. 'When two of the sides have been mentioned, the remaining side is often called the base.' Well, but how if two of the sides have *not* been mentioned?

Nos. In that case we do not use the word.

Min. Do you not? Turn to p. 22, Th. 2, Cor. 1, 'Triangles on the same or equal bases and of equal altitude are equal.'

Nos. We abandon the point.

Min. You had better abandon the Definition.

P. 12, Def. 34. Is not 'identically equal' tautology? Things that are 'identical' must surely be 'equal' also. Again, '*every part* of one being equal,' &c. What do you mean by 'every part' of a rectilineal figure?

Nos. Its sides and angles, of course.

Min. Then what do you mean by *Ax* (*b*) in p. 3. 'The whole is equal to the sum of its parts'? This time, I think I need not 'pause for a reply'!

P. 15, Def. 38. 'When a straight Line intersects two other straight Lines it makes with them eight angles etc.'

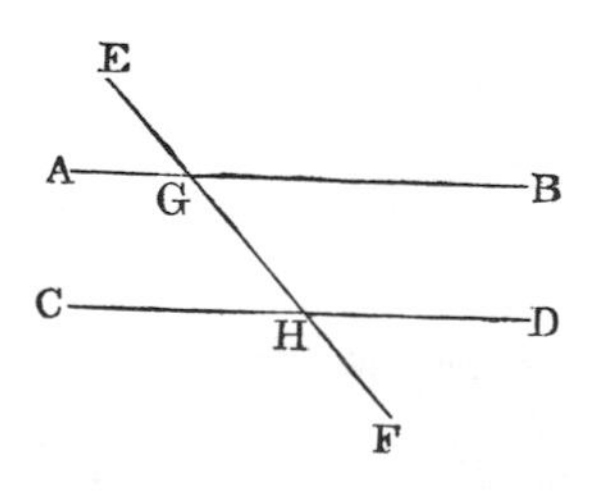

Let us count the angles at *G*. They are, the 'major'

and 'minor' angles which bear the name *EGA*; do. for *EGB*; do. for *AGH*; and do. for *BGH*. That is, eight angles at *G* alone. There are sixteen altogether.

P. 17, Th. 30. 'If a quadrilateral has two opposite sides equal and parallel, it is a Parallelogram.'

This re-asserts part of its own data.

P. 17, Th. 31. 'Straight Lines that are equal and parallel have equal projections on any other straight Line; conversely, parallel straight Lines that have equal projections on another straight Line are equal.'

The first clause omits the case of Lines that are equal and in one and the same straight Line. The second clause

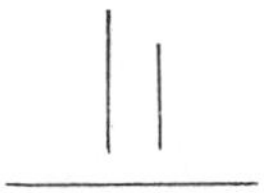

is not true: if the parallel Lines are at right angles to the other Line, their projections are equal, both being zero, whether the Lines are equal or not.

P. 18, Th. 32. 'If there are three parallel straight Lines, and the intercepts made by them on any straight Line that cuts them are equal, then etc.'

The subject of this Proposition is inconceivable: there are *three* intercepts, and by no possibility can these three be equal.

P. 25, Prob. 5. 'To construct a rectilineal Figure equal to a given rectilineal Figure and having the number of its sides one less than that of the given Figure.'

May I ask you to furnish me with the solution of this

Problem, taking, as your 'given rectilineal Figure,' a Triangle?

Nos. (*indignantly*) I decline to attempt it!

Min. I will now sum up the conclusions I have come to with respect to your Syllabus.

In the subjects of Lines, Angles, and Parallels, the changes you propose are as follows:—

You give a very unsatisfactory Definition of a 'Right Line,' and then most illogically re-state it as an Axiom.

You extend the Definition of Angle—a most disastrous innovation.

Your Definition of 'Right Angle' is a failure.

You substitute Playfair's axiom for Euclid's 12th.

All these things are very poor compensation indeed for the vital changes you propose—the separation of Problems and Theorems, and the abandonment of Euclid's order and numeration. Restore the Problems (which are also Theorems) to their proper places, keep to Euclid's numbering (interpolating your new Propositions where you please), and your Syllabus may yet prove to be a valuable addition to the literature of Elementary Geometry.

ACT III.

SCENE II.

§ 2. WILSON'S 'SYLLABUS'-MANUAL.

> 'No followers allowed.'
>
> TIMES' ADVERTISEMENT-SHEET, *passim*.

Nie. I lay before you '*Elementary Geometry, following the Syllabus prepared by the Geometrical Association,*' by J. M. WILSON, M.A., 1878.'

Min. In what respects is this book a 'Rival' of Euclid?

Nie. Well, it separates Problems from Theorems ——

Min. Already discussed (see p. 18).

Nie. It adopts Playfair's Axiom ——

Min. Discussed (see p. 40).

Nie. It abandons diagonals in Book II ——

Min. Discussed (see p. 50).

Nie. And it adopts a new sequence and numeration.

Min. *That*, of course, prevents us from taking it as merely a new edition of Euclid. It will need very strong evidence indeed to justify its claim to set aside the sequence and numeration of our old friend. We must now examine the book *seriatim.* When we come to matters that have been already condemned, either in Mr. Wilson's book, or in the 'Syllabus,' I shall simply note the fact. We need have no new discussion, except as to new matter.

Nie. Quite so.

Min. In the 'Introduction,' at p. 2, I read '*A Theorem is the formal statement of a Proposition,*' &c. Discussed at p. 189.

At p. 3 we have the 'Rule of Conversion,' which I have already endeavoured to understand (see p. 190).

At p. 6 is a really remarkable assertion. '*Every Theorem may be shewn to be a means of indirectly measuring some magnitude.*' Kindly illustrate this on Euc. I. 14.

Nie. (*hastily*) Oh, if you pick out one single accidental excep ——

Min. Well, then, take 16, if you like: or 17, or 18 ——

Nie. Enough, enough!

Min. (*raising his voice*)—or 19, or 20, or 21, or 24, or 25, or 27, or 28, or 30!

Nie. We abandon '*every.*'

Min. Good. At p. 8 we have the Definitions of '*major conjugate*' and '*minor conjugate*' (discussed at p. 185).

At p. 9 is our old friend the '*straight angle*' (see p. 101).

In the same page we have that wonderful triad of Lines, one of which is '*regarded as lying between the other two*' (see p. 185).

And also the extraordinary result that follows when one straight Line '*stands upon another*' (see p. 186).

At p. 27, Theorem 14, is a new proof of Euc. I. 24, apparently an amended version of Mr. Wilson's five-case proof, which I discussed at p. 137. He has now reduced it to three cases, but I still think the '*bisector of the angle*' a superfluity.

At p. 37 we have those curious specimens of '*Theorems of Equality*,' which I discussed at p. 139.

At p. 53 is the Theorem which asserts, in its conclusion, part of its own *data* (see p. 192).

At p. 54 we are told that '*parallel Lines, which have equal projections on another Line, are equal*' (see p. 193).

At p. 55 we have the inconceivable triad of '*equal intercepts*' made by a Line cutting three Parallels (see p. 193).

At p. 161 I am surprised to see him fall into a trap in which I have often seen unwary students caught, while trying to say Euc. III. 30 ('To bisect a given arc') After proving two chords equal, they at once conclude that *certain* arcs, cut off by them, are equal; forgetting to prove that the arcs in question are both *minor* arcs.

But I must go no further: I have already wandered beyond the limits of Euc. I, II. The one great merit of this book ——

Nie. You have mentioned all the *faults*, then?

Min. By no means. You are too impatient. The one great merit, as I was saying, of Mr. Wilson's new book (and a most blessed change it is!) is that it ignores the whole theory of 'direction.' That he has finally abandoned that night-mare of Elementary Geometry, I dare not hope: so all I have said about it had better stand, lest in some future fit of inspiration he should bring out a yet more agonising version of it.

But it has the usual hiatus of a system which replaces Euclid's Axiom by Playfair's: it provides no means of proving that the Lines contemplated by Euclid will meet if produced. (This I have discussed at p. 187.)

Its proposed changes in the *sequence* of Euclid I have discussed at p. 188.

It has a few other faults, which I have already discussed in Mr. Wilson's own book, and a few peculiar to the Syllabus; but I spare you such minute criticisms.

But what I have now to ask you is simply this. What possible pretext have you left for suggesting that Euclid's Manual, and specially his sequence and numeration, should be abandoned in favour of this far from satisfactory infant?

Nie. There are some new Theorems ——

Min. *Those* constitute no reason: you might easily interpolate them.

Nie. I fear there are no other grounds to urge. But I should like to consult the *doppelgänger* of the Association before I throw up my brief.

Min. By all means.

[*For a minute or two there is heard a rustling and a whispering, as of ghosts. Then* NIEMAND *speaks again.*]

Nie. They think that, considering that this book is but just published, and that it is definitely put forward as *the* Manual to supersede Euclid, it ought to be examined more in detail, with reference to what is *new* in it—that is, new proofs of Euclid's Propositions, and new Propositions.

Min. (*with a weary sigh*) Very well. It will perhaps be more satisfactory to do this, if only to ascertain exactly how much this new Manual contains that is really new and really worthy of adoption. But I shall limit my examination to the subject-matter of Euc. I, II.

Nie. That is all we ask.

Min. We begin, then, at p. 12.

Theorem 1. '*All right angles are equal.*' This is proved by their being halves of a 'straight angle,' a phrase which I have already criticised. There is a rather important omission in the proof, no distinction being drawn between the 'straight angle' on one side of a Line, and the other (of course named by the same letters) which lies on the other side and completes the four right angles. This Theorem, if proved without 'straight angles,' might be worth adding to a new edition of Euclid.

Th. 2 (p. 13) is Euc. I. 13, proved as in Euclid.

Th. 3 (p. 14) is Euc. I. 14, where, unfortunately, a new proof is attempted, which involves a fallacy. It is deduced from an 'Observation' in p. 9, that 'a straight Line makes with its continuation at any point an angle of two right angles,' which deduction can be effected only by the process of converting a universal affirmative '*simpliciter*' instead of '*per accidens.*'

Th. 4 (p. 14) is Euc. I. 15, proved as in Euclid.

At p. 17 I find a 'Question.' '*State the fact that "all geese have two legs" in the form of a Theorem.*' This I would not mind attempting; but, when I read the additional request, to '*write down its converse theorem,*' it is so powerfully borne in upon me that the writer of the Question is probably himself a biped, that I feel I must, however reluctantly, decline the task.

Th. 5 (p. 18) is Euc. I. 4, proved as in Euclid.

Th. 6 (p. 20) is Euc. I. 5, proved by supposing the vertical angle to be bisected, thus introducing a 'hypothetical construction' (see p. 20).

Th. 7 (p. 21) is Euc. I. 26 (1st part), proved by superposition. Euclid's proof, by making a new Triangle, is quite as good, I think. The areas are here proved to be equal, a point omitted by Euclid: I think it a desirable addition to the Theorem.

Th. 8 (p. 22) is Euc. I. 5, proved by reversing the Triangle and then placing it *on itself* (or on the trace it has left behind), a most objectionable method (see p. 48).

Theorems 9 to 13 (pp. 22 to 26) are Euc. I. 16, 18, 19, 20, 21, with Euclid's proofs.

Th. 14 (p. 27) is Euc. I. 24, proved by supposing an angle to be bisected: another 'hypothetical construction.'

Th. 15 (p. 28) is Euc. I. 8, for which two proofs are offered:—one by Euc. I. 24 (which seems to be reversing the natural order)—the other by an application of Euc. I. 5, a method involving *three* cases, of which only one is given. All this is to save the introduction of Euc. I. 7, a Theorem which *I* think should by no means be omitted. (See p. 220.) Here, as in Th. 7, the equality of the areas is, I think, a desirable addition to Euclid's Theorem.

Th. 16 (p. 29) is Euc. I. 25, with old proof.

Th. 17 (p. 30) is Euc. I. 26 (2nd part) proved by superposition instead of Euclid's method (which I prefer) of constructing a new Triangle.

Th. 18 (p. 32) is Euc. I. 17, with old proof.

Th. 19 (p. 33) is new. '*Of all the straight Lines that can be drawn from a given point to meet a given straight Line, the perpendicular is the shortest; and of the others, those making equal angles with the perpendicular are equal; and that which makes a greater angle with the perpendicular is greater than that which makes a less.*' This I think deserves to be interpolated.

Th. 20 (p. 34) is new. '*If two Triangles have two sides of the one equal to two sides of the other, each to each, and the angles opposite to two equal sides equal, the angles opposite to the other two equal sides are either equal or supplementary, and in the former case the Triangles are equal in all respects.*' I do not think it worth while to trouble a beginner with this rather obscure Theorem, which is of no practical use till he enters on Trigonometry.

Th. 21 (p. 43) is Euc. I. 27: old proof.

Th. 22 (p. 44) is Euc. I. 29 (1st part), proved by Euc. I. 27 and Playfair's Axiom (see p. 40).

Th. 23 (p. 45) is new. '*If a straight Line intersects two other straight Lines and makes either a pair of alternate angles equal, or a pair of corresponding angles equal, or a pair of interior angles on the same side supplementary; then, in each case, the two pairs of alternate angles are equal, and the four pairs of corresponding angles are equal, and the two pairs of interior angles on the same side are supplementary.*' This most formidable enunciation melts down into the mildest proportions when superfluities are omitted, and only so much of it proved as is really necessary to include the whole. Euclid proves all that is valuable in it in the course of I. 29, and I do not see any sufficient reason for stating and proving it as a separate Theorem.

Th. 23, Cor. (p. 46) is the rest of Euc. I. 29: old proof.

Th. 24 (p. 46) is Euc. I. 30, proved as a Contranominal of Playfair's Axiom.

Th. 25, 26, and Cor. (pp. 47, 48) are Euc. I. 32 and Corollaries: old proof.

Th. 27, 1st part (p. 50), is a needless repetition of part of the Corollary to Th. 23.

Th. 27, 2nd part (p. 50), is part of Euc. I. 34: old proof.

Th. 28 (p. 51) is the rest of Euc. I. 34, proved as in Euclid.

Th. 29 (p. 52) is new. '*If two Parallelograms have two adjacent sides of the one respectively equal to two adjacent sides of the other, and likewise an angle of the one equal to an angle of the other; the Parallelograms are identically equal.*' This might be a useful *exercise* to set; but really it does not seem of sufficient importance to be selected for a Manual.

Th. 30 (p. 53) is Euc. I. 33: old proof.

Th. 31 (p. 54) is new. '*Straight Lines which are equal and parallel have equal projections on any other straight Line; conversely, parallel straight Lines which have equal projections on another straight Line are equal; and equal straight Lines, which have equal projections on another straight Line, are equally inclined to that Line.*' The first and third clauses might be interpolated, though I think their value doubtful. The second is false. (See p. 193.)

Th. 32 (p. 55) is new. '*If there are three parallel straight Lines, and the intercepts made by them on any straight Line that cuts them are equal, then the intercepts on any other straight Line that cuts them are equal.*' This is awkwardly worded (in fact, as it stands, its subject, as I pointed out in p. 193, is inconceivable), and does not seem at all worth stating as a Theorem.

At p. 57 I see an 'Exercise' (No. 5). '*Shew that the angles of an equiangular Triangle are equal to two-thirds of a right angle.*' In this attempt I feel sure I should fail. In early life I was taught to believe them equal to *two right angles*—an antiquated prejudice, no doubt; but it is difficult to eradicate these childish instincts.

Problem 1 (p. 61) is Euc. I. 9: old proof. It provides no means of finding a radius 'greater than half *AB*,' which would seem to require the previous bisection of *AB*. Thus the proof involves the fallacy '*Petitio Principii.*'

Pr. 2 (p. 62) is Euc. I. 11, proved nearly as in Euclid.

Pr. 3 (p. 62) is Euc. I. 12, proved nearly as in Euclid. It omits to say how a 'sufficient radius' can be secured, a point not neglected by Euclid.

Pr. 4 (p. 63) is Euc. I. 10, proved nearly as in Euclid. This also, like Pr. 1, involves the fallacy '*Petitio Principii.*'

Pr. 5 (p. 64) is Euc. I. 32, proved nearly as in Euclid, but claims to use compasses to transfer distances, a Postulate which Euclid has (properly, I think) treated as a Problem. (See p. 212.)

Pr. 6, 7 (pp. 65, 66) are Euc. I. 23, 31: old proofs.

Problems 8 to 11 (pp. 66 to 69) are new. Their object is to construct Triangles with various *data*: viz. A, B, and c; A, B, and a; a, b, and C; a, b, and A. They are good exercises, I think, but hardly worth interpolating as Theorems. The first of them is remarkable as one of the instances where Mr. Wilson assumes Euc. Ax. 12, without giving, or even suggesting, any proof. If he intends to assume it as an *Axiom*, he makes Playfair's Axiom superfluous. No Manual ought to assume *both* of them.

Theorem 1 (p. 82) is Euc. I. 35, proved as in Euclid, but incompletely, as it only treats of one out of three possible cases.

Th. 2 (p. 83) is new. '*The area of a Triangle is half the area of a rectangle whose base and altitude are equal to those of the Triangle.*' This is merely a particular case of Euc. I. 41, and may fairly be reserved till we enter on Trigonometry, where it first begins to have any practical value.

Th. 2, Cor. 1 (p. 84) is Euc. I. 37, 38: old proofs.

Th. 2, Cor. 2 (p. 84) is new. '*Equal Triangles on the same or equal bases have equal altitudes.*' No proof is offered. It is an easy deduction, of questionable value.

Th. 2, Cor. 3 (p. 84) is Euc. I. 39, 40. No proof given.

Th. 3 (p. 84) is new. '*The area of a trapezium* [by which Mr. Wilson means '*a quadrilateral that has only one pair of opposite sides parallel*'] *is equal to the area of a rectangle whose base is half the sum of the two parallel sides, and whose altitude is the perpendicular distance between them.*' I have no hesitation in pronouncing this to be a mere 'fancy' Proposition, of no practical value whatever.

Th. 4 (p. 86) is Euc. I. 43: old proof.

Th. 5 (p. 87) is Euc. II. 1 : old proof.

Th. 6, 7, 8 (p. 88, &c.) are Euc. II. 4, 7, 5. The sequence of Euc. II. 5, and its Corollary, is here inverted. Also the diagonals are omitted, and nearly every detail is left unproved, thus attaining a charming brevity—of *appearance!*

Th. 9 (p. 91) is Euc. I. 47 : old proof.

Th. 10, 11 (pp. 94, 95) are Euc. 12, 13 : old proof.

Th. 12 (p. 95) is new. '*The sum of the squares on two sides of a Triangle is double the sum of the squares on half the base and on the line joining the vertex to the middle point of the base.*' This, Mr. Wilson tells us, is 'Apollonius' Theorem': but, even with that mighty name to recommend it, I cannot help thinking it rather more curious than useful.

Th. 13 (p. 96) is Euc. II. 9, 10. Proved algebraically, and thus degraded from the position of a (fairly useful) geometrical Theorem to a mere addition-sum, of no more value than millions of others like it.

In the next proposition we suddenly transfer our allegiance, for no obvious reason, from Arabic to Latin numerals.

Problem I (p. 99) is Euc. I. 42 : old proof.

Pr. II. (p. 100) is Euc. I. 44: proved nearly as in Euclid, but labours under the same defect as Pr. 8 (p. 66) in that it assumes, without proof, Euc. Ax. 12.

Pr. III (p. 100) is Euc. I. 45 : old proof.

Pr. IV (p. 101) is Euc. II. 14 : old proof.

Pr. V (p. 103) is new. '*To construct a rectilineal Figure equal to a given rectilineal Figure and having the number of its sides one less than that of the given figure ; and thence to construct a Triangle equal to a given rectilineal Figure.'* This I have already noticed (see p. 193). It really is not worth interpolating as a new Proposition. And its concluding clause is, if I may venture on so harsh an expression, childish: it reminds me of nothing so much as the Irish patent process for making cheap shoes—by taking boots and cutting off the tops!

Pr. VI (p. 103) is '*To divide a straight Line, either internally or externally, into two segments such that the rectangle contained by the given Line and one of the segments may be equal to the square on the other segment.*' The case of *internal* section is Euc. II. 11, with the old proof. The other case is new, and worth interpolating.

I have now discussed, with as much care and patience as the lateness of the hour will permit, so much of this new Manual as corresponds to Euc. I, II, and I hope your friends are satisfied.

[*A gentle cooing, as of satisfied ghosts, is heard in the air.*]

I will now give you in a few words the net result of it all, and will show you how miserably small is the basis on which Mr. Wilson and his coadjutors of the 'Association' rest their claim to supersede the Manual of Euclid.

[*An angry moaning, as of ghosts suffering from neuralgia, surges round the room, till it dies away in the chimney.*]

By breaking up certain of the Propositions of Euc. I, II, and including some of the Corollaries, we get 73 Propositions in all—57 Theorems and 16 Problems. Of these 73, this Manual omits 14 (10 Theorems and 4 Problems); it proves 43 (32 Theorems and 11 Problems) by methods almost identical with Euclid's; for 10 of them (9 Theorems and a Problem) it offers new proofs, against which I have recorded my protest, one being illogical, 2 (needlessly) employing 'superposition,' 2 deserting Geometry for Algebra, and the remaining 4 omitting the diagonals in Euc. II; and finally it offers 6 new proofs, which I think may fairly be introduced as alternatives for those of Euclid.

In all this, and in all the matters previously discussed, I fail to see one atom of reason for abandoning Euclid. Have you any yet-unconsidered objections to urge against my proposal 'that the sequence and numeration of Euclid be kept unaltered'?

[*Dead silence is the only reply.*]

Carried, *nemine contragemente!* And now, Prisoner at the Bar (I beg your pardon, I should say 'Professor on the Sofa'), have you, and your attendant phantoms, any other reasons to urge for regarding this Manual as in any sense a substitute for Euclid's—as in any sense anything else than a revised edition of Euclid?

Nie. We have nothing more to say.

Min. Then I can but repeat with regard to this newborn 'follower' of the Syllabus, what I said of the

Syllabus itself. Restore the Problems (which are also Theorems) to their proper places; keep to Euclid's numbering (interpolating your new Propositions where you please); and your new book may yet prove a valuable addition to the literature of Elementary Geometry.

[*A tremulous movement is seen amid the ghostly throng. They waver fitfully to and fro, and finally drift off in the direction of one corner of the ceiling. When the procession has got well under way,* NIEMAND *himself becomes hazy, and floats off to join them. The whole procession gradually melts away into vacancy,* DIAMOND *going last, nibbling at the heels of* NERO, *for which a pair of gorgeous Roman sandals seem to afford but scanty protection.*]

ACT IV.

'Old friends are best.'

[*Scene as before. Time, the early dawn.* MINOS *slumbering uneasily, having fallen forwards upon the table, his forehead resting on the inkstand. To him enter* EUCLID *on tip-toe, followed by the phantasms of* ARCHIMEDES, PYTHAGORAS, ARISTOTLE, PLATO, *&c., who have come to see fair play.*]

§ 1. *Treatment of Pairs of Lines.*

Euc. Are all gone?

Min. 'Be cheerful, sir:
Our revels now are ended: these our actors,
As I foretold you, were all spirits, and
Are melted into air, into thin air!'

Euc. Good. Let us to business. And first, have you found any method of treating Parallels to supersede mine?

Min. No! A thousand times, no! The infinitesimal method, so gracefully employed by M. Legendre, is unsuited to beginners: the method by transversals, and the method by revolving Lines, have not yet been offered in a logical form: the 'equidistant' method is too cumbrous:

and as for the method of 'direction,' it is simply a rope of sand—it breaks to pieces wherever you touch it!

Euc. We may take it as a settled thing, then, that you have found no sufficient cause for abandoning either my sequence of Propositions or their numbering, and that all that now remains to be considered is whether any important modifications of my Manual are desirable?

Min. Most certainly.

Euc. Have you met with any striking novelty on the subject of a practical test for the meeting of Lines?

Min. There is *one* rival to your 12th Axiom which is formidable on account of the number of its advocates—the one usually called 'Playfair's Axiom.'

Euc. We have discussed that matter already (p 40).

Min. But what have you to say to those who reject Playfair's Axiom as well as yours?

Euc. I simply ask them what practical test, as to the meeting of two given finite Lines, they propose to employ. Not only will they find it necessary to prove, in certain Theorems, that two given finite Lines will meet if produced, but they will even find themselves sometimes obliged to prove it of two Lines, of which the only geometrical fact known is that they possess the very property which forms the subject of my Axiom. I ask them, in short, this question:—'Given two Lines making, with a certain transversal, two interior angles together less than two right angles, how do you propose to prove, without my Axiom, that they will meet if produced?'

Min. The advocates of the 'direction' theory would of course reply, 'We can prove, from the given property, that they have different directions: and then we bring in the Axiom that Lines having different directions will meet if produced.'

Euc. All *that* you have satisfactorily disposed of in your review of Mr. Wilson's Manual.

Min. The only other substitute, that I know of, belongs to the 'equidistant' theory, which replaces your Axiom by three or four new Axioms and six new Theorems. *That* substitute, also, I have seen reason to reject.

My general conclusion is that your method of treatment of all these subjects is the best that has yet been suggested.

Euc. Any noticeable innovations in the treatment of Right Lines and Angles?

Min. Those subjects I should be glad to talk over with you.

Euc. With all my heart. And now how do you propose to conduct this our final interview?

Min. I should wish, in the first place, to lay before you the general charges which have been brought against you: then to discuss your treatment of Lines and Angles, as contrasted with that of your 'Rivals'; and lastly the omissions, alterations, and additions proposed by them.

Euc. Good. Let us begin.

Min. I will take the general charges under three headings:—Construction, Demonstration, and Style. And first as to Construction:—

§ 2. *Euclid's Constructions.*

I am told that you indulge too much in 'arbitrary restrictions.' Mr. Reynolds says (Pref. p. vi.) 'The arbitrary restrictions of Euclid involve him in various inconsistencies, and exclude his constructions from use. When, for instance, in order to mark off a length upon a straight Line, he requires us to describe five Circles, an equilateral Triangle, one straight line of limited, and two of unlimited length, he condemns his system to a divorce from practice at once and from sound reason.'

Euc. Mr. Reynolds has misunderstood me: I do not require all that construction in Prop. 3. To explain my meaning I must go back to Prop. 2, and I must ask your patience while I make a few general remarks on construction. The machinery I allow consists of a pencil, a ruler, and a pair of compasses to be used for drawing a Circle about a given centre and *passing through a given point* (that is what I mean by 'at any distance'), but *not* to be used for transferring distances from one part of a diagram to another *until it has been shown that such transference can be effected by the machinery already allowed.*

Min. But why not allow such transference without proving that possibility?

Euc. Because it would be introducing as a *Postulate* what is really a *Problem.* And I go on the general principle of never putting a Problem among my Postulates, nor a Theorem among my Axioms.

Min. I heartily agree in your general principle, though I need scarcely remind you that it has been frequently charged against you, as a *fault*, that you state as an Axiom what is really a Theorem.

Euc. That charge has been met (see p. 40). To return to my subject. I merely prove, once for all, in Prop. 2, that a Line *can* be drawn, from a given point, and equal to a given Line, by the original machinery alone, and *without* transferring distances. After that, my reader is welcome to transfer a distance by any method that comes handy, such as a bit of string &c.: and of course he may now transfer his compasses to a new centre. And this is all I expect him to do in Prop. 3.

Min. Then you *don't* expect these five Circles &c. to be drawn whenever we have to cut off, from one Line, a part equal to another?

Euc. *Pas si bête, mon ami.*

Min. Some of your Modern Rivals are, however, a little discontented with the very scanty machinery you allow.

Euc. 'A bad workman always quarrels with his tools.'

Min. Their charge against you is 'the exclusion of hypothetical constructions.' Mr. Wilson says (Pref. p. i.) 'The exclusion of hypothetical constructions may be mentioned as a self-imposed restriction which has made the confused order of his first book necessary, without any compensating advantage.'

Euc. In reply, I cannot do better than refer you to Mr. Todhunter's Essay on Elementary Geometry (p. 186). 'Confused order is rather a contradictory expression,' &c. (see p. 241).

Min. Your reply is satisfactory. Mr. Wilson himself is an instance of the danger of such a method. Three times at least (pp. 46, 70, 88) he produces Lines to meet without attempting to prove that they *will* meet.

§ 3. *Euclid's Demonstrations.*

Min. The next heading is 'Demonstration.' You are charged with an 'invariably syllogistic form of reasoning.' (Wilson, Pref. p. i.)

Euc. Do you know, I am vain enough to think that a merit rather than a defect? Let me quote what Mr. Cuthbertson says on this point (Pref. p. vii.). 'Euclid's mode of demonstration, in which the conclusion of each step is preceded by reasoning expressed with all the exactness of the minor premiss of a syllogism, of which some previous proposition is the major premiss, has been adopted as offering a good logical training, and also as being peculiarly adapted for teaching large classes, rendering it possible for the teacher to call first upon one, then upon another, and so on, to take up any link in the chain of argument.' Perhaps even Mr. Wilson's own book would not be the worse if the reasoning were a *trifle* more 'syllogistic'!

Min. A fair retort. You are also charged with 'too great length of demonstration.' Mr. Wilson says (Pref. p. i.) 'The real objections to Euclid as a text-book are . . . the length of his demonstrations.' And Mr. Cooley says (Pref. p. 1.) 'The important and fertile theorems, which crown the heights in this field of knowledge, are here all retained, and those only are omitted which seem to be but the steps of a needlessly protracted ascent. The short road thus opened will be found perfectly solid in construction, and at the same time far less tedious and fatiguing than the circuitous one hitherto in vogue.'

Euc. I think Mr. Wilson's Th. 17 (p. 27), with its five figures (all necessary, though he only draws one), and still more his marvellous Problem, 'approached by four stages,' which fills pages 69 to 72, are pretty good instances of lengthy demonstration. And Mr. Cooley's 'short and solid road' contains, if I remember right, a rather break-neck crevasse!

Min. The next charge against you is 'too great *brevity* of demonstration.' Mr. Leslie (a writer whom I have not thought it necessary to review as a 'Modern Rival,' as his book is nearly seventy years old) says (Pref. p. vi.) 'In adapting it' (the Elements of Euclid) 'to the actual state of the science, I have . . . sought to enlarge the basis . . . The numerous additions which are incorporated in the text, so far from retarding will rather facilitate progress, by rendering more continuous the chain of demonstration. To multiply the steps of ascent, is in general the most expeditious mode of gaining a summit.'

Euc. I think you had better refer him to Mr. Wilson and Mr. Cooley: they will answer *him*, and he in his turn will confute *them*!

Min. The last charge relating to demonstration is, in Mr. Wilson's words (Pref. p. viii.) 'the constant reference to general Axioms and general Propositions, which are no clearer in the general statement than they are in the particular instance,' which practice, he says, makes the study of Geometry 'unnecessarily stiff, obscure, tedious and barren.'

Euc. *One* advantage of making a general statement, and afterwards referring to it instead of repeating it, is that you have to go through the mental process of affirming or proving the truth *once for all*: apparently Mr. Wilson would have you begin *de novo* and think out the truth every time you need it! But the great reason for always referring back to your universal, instead of affirming the particular (Mr. Wilson is merely starting the old logical hare 'Is the syllogism a *Petitio Principii*?'), is that the truth of the particular does not rest on any data peculiar to itself, but on general principles applicable to all similar cases; and that, *unless those general principles prove the conclusion for* all *cases, they cannot be warranted to prove it for any one selected case.* If, for instance, I see a hundred men, and am told that some assertion is true of ninety-nine of them, but am *not* told that it is true of *all*, I am not justified in affirming it of any selected man; for he *might* chance to be the excepted one. Now the assertion, that the truth of the particular case under notice depends on general principles, and not on peculiar circumstances, is neither more nor less than the assertion of the *universal* affirmative which Mr. Wilson deprecates.

§ 4. *Euclid's Style.*

Min. Quite satisfactory. I will now take the third heading, namely 'Style.'

You are charged with Artificiality, Unsuggestiveness, and Want of Simplicity. Mr. Wilson says (Pref. p. i.) 'The real objections to Euclid as a text-book are his artificiality . . . and his unsuggestiveness,' and again, 'he has sacrificed, to a great extent, simplicity and naturalness in his demonstrations, without any corresponding gain in grasp or cogency.'

Euc. Well, really I cannot deal with general charges like these. I prefer to abide by the verdict of my readers during these two thousand years. As to 'unsuggestiveness,' that is a charge which cannot, I admit, be retorted on Mr. Wilson: his book is *very* suggestive—of remarks which, perhaps, would not be *wholly* 'music to his ear'!

§ 5. *Euclid's treatment of Lines and Angles.*

Min. Let us now take the subjects of Right Lines and Angles; and first, the 'Right Line.'

I see, by reference to the original, that you define it as a Line 'which lies evenly as to points on it.' That of course is only an attempt to give the mind a grasp of the idea. It leads to no geometrical results, I think?

Euc. No: nor does *any* definition of it, that I have yet seen.

Min. I have no rival Definitions to propose. Mr. Wilson's 'which has the same direction at all parts of its length' has perished in the collapse of the 'direction' theory: and M. Legendre's 'the shortest course from one point to another' is not adapted for the use of a beginner.

And I do not know that any change has been suggested in your test of a right Line in Prop. 14.

The next subject is 'Angles.'

Your definition would perhaps be improved, if for 'inclination to' we were to read 'declination from,' for, the greater the angle the greater the *de*clination, and the less (as it seems to me) the *in*clination.

Euc. I agree with you.

Min. The next point is that you limit the size of an angle to something less than the sum of two right angles.

Euc. What advantage is claimed for the extension of the Definition?

Min. It is a prospective rather than an immediate one. It must be granted you that the larger angles are not needed in the first four Books—

Euc. In the first *six* Books.

Min. Nay, surely you need them in the Sixth Book?

Euc. Where?

Min. In Prop. 33, where you treat of 'any equimultiples whatever' of an angle, of an arc, and of a sector. You cannot possibly assume the multiple angle to be always less than two right angles.

Euc. You think, then, that a multiple of an angle must itself be an angle?

Min. Surely.

Euc. Then a multiple of a man must itself be a man. If I contemplate a man as multiplied by the number ten thousand, I must realise the idea of a man ten thousand times the size of the first?

Min. No, you need not do *that*.

Euc. Thanks: it *is* rather a strain on the imaginative faculty.

Min. You mean, then, that the multiple of an angle may be conceived of as so many separate angles, not in contact, nor added together into one?

Euc. Certainly.

Min. But you have to contemplate the case where two such angular magnitudes are equal, and to infer from that, by III. 26, that the subtending arcs are equal. How can you infer this when your angular magnitude is not one angle but many?

Euc. Why, the sum total of the first set of angles is equal to the sum total of the second set. Hence the second set can clearly be broken up and put together again in such amounts as to make a set equal, each to each, to the first set: and then the sum total of the arcs, and likewise of the sectors, will evidently be equal also.

But if you contemplate the multiples of the angles as single angular magnitudes, I do not see how you prove the equality of the subtending arcs: for *my* proof applies only to cases where the angle is less than the sum of two right angles.

Min. That is very true, and you have quite convinced me that we ought to observe that limit, and not contemplate 'angles of rotation' till we enter on the subject of Trigonometry.

As to right angles, it has been suggested that your Axiom 'all right angles are equal to one another' is capable of proof as a Theorem.

Euc. I do not object to the interpolation of such a

Theorem, though there is very little to distinguish so simple a Theorem from an Axiom.

Min. Let us now consider the omissions, alterations, and additions, which have been proposed by your Modern Rivals.

§ 6. *Omissions, alterations, and additions, suggested by Modern Rivals.*

Euc. Which of my Theorems have my Modern Rivals proposed to omit?

Min. Without dwelling on such extreme cases as that of Mr. Pierce, who omits no less than 19 of the 35 Theorems in your First Book, I may say that the only two, as to which I have found anything like unanimity, are I. 7 and II. 8.

Euc. As to I. 7, I have several reasons to urge in favour of retaining it.

First, it is useful in proving I. 8, which, without it, is necessarily much lengthened, as it then has to include *three* cases: so that its omission effects little or no saving of space.

Secondly, the modern method of proving I. 8 independently leaves I. 7 still unproved.

Min. *That* reason has no weight unless you can prove I. 7 to be valuable for itself.

Euc. True, but I think I *can* prove it; for, thirdly, it shows that, of all plane Figures that can be made by hingeing rods together, the *three*-sided ones (and these only) are *rigid* (which is another way of stating the fact that there cannot be *two* such figures on the same base). This is analogous to the fact, in relation to solids contained by plane surfaces hinged together, that *any* such solid is rigid, there being no maximum number of sides.

And fourthly, there is a close analogy between I. 7, 8 and III. 23, 24. These analogies give to Geometry much of its beauty, and I think that they ought not to be lost sight of.

Min. You have made out a good case. Allow me to contribute a 'fifthly.' It is one of the very few Propositions that have a direct bearing on practical science. I have often found pupils much interested in learning that the principle of the rigidity of Triangles is of constant use in architecture, and even in so homely a matter as the making of a gate.

The other Theorem which I mentioned, II. 8, is now so constantly ignored in examinations that it is very often omitted, as a matter of course, by students. It is believed to be extremely difficult and entirely useless.

Euc. Its difficulty has, I think, been exaggerated. Have you tried to teach it?

Min. I *have* occasionally found pupils amiable enough to listen to what they felt sure would be of no service in examinations. My experience has been wholly among undergraduates, any one of whom, if of average ability, would, I think, master it in from five to ten minutes.

Euc. No very exorbitant demand on your pupil's time. As to its being 'entirely useless,' I grant you it is of no *immediate* service, but you will find it eminently useful when you come to treat the Parabola geometrically.

Min. That is true.

Euc. Let us now consider the new methods of proof suggested by my Rivals.

Min. Prop. 5 has been much attacked—I may say trampled on—by your Modern Rivals.

Euc. Good. So that is why you call it 'The Asses' Bridge'? Well, how many new methods do they suggest for crossing it?

Min. One is 'hypothetical construction,' M. Legendre bisecting the base, and Mr. Pierce the vertical angle, but without any proof that the thing can be done.

Euc. So long as we agree that beginners in Geometry shall be limited to the use of Lines and Circles, so long will it be unsafe to assume a point as found, or a Line as drawn, merely because we are sure it *exists*. For example, it is axiomatic, of course, that every angle has a bisector: but it is equally obvious that it has two trisectors: and if I may assume the one as drawn, why not the others also? However we have discussed this matter already (p. 20).

Min. A second method is 'superposition,' adopted by Mr. Wilson and Mr. Cuthbertson—a method which here involves the *reversing* of the triangle, before applying it to its former position.

Euc. That also we have discussed (p. 47). What is the method adopted in the new Manual founded on the Syllabus of the Association?

Min. The same as Mr. Pierce's. Mr. Reynolds has a curious method: he treats the sides as obliques 'equally remote from the perpendicular.'

Euc. Curious, indeed.

Min. But perhaps the most curious of all is Mr. Willock's method: *he* treats the sides as radii of a circle, and the base as a chord.

Euc. He had better have made them asymptotes of a hyperbola at once! *C'est magnifique, mais ce n'est pas la—Géométrie.*

Min. Two of your Rivals prove Prop. 8 from Prop. 24.

Euc. 'Putting the cart before the horse,' in my humble opinion.

Min. For a brief proof of Prop. 13, let me commend to your notice Mr. Reynolds'—consisting of the seven words 'For they fill exactly the same space.'

Euc. Why so lengthy? The word 'exactly' is superfluous.

Min. Instead of your chain of Theorems, 18, 19, 20, several writers suggest 20, 19, 18, making 20 axiomatic.

Euc. That has been discussed already (p. 56).

Min. Mr. Cuthbertson's proof of Prop. 24 is, if I may venture to say so, more complete than yours. He constructs his diagram without considering the lengths of the sides, and then proves the 3 possible cases separately.

Euc. I think it an improvement.

Min. There are no other noticeable innovations, that have not been already discussed, except that Mr. Cuthbertson proves a good deal of Book II by a quasi-algebraical method, without exhibiting to the eye the actual Squares and Rectangle: while Mr. Reynolds does it by pure algebra.

Euc. I think the actual Squares, &c. most useful for beginners, making the Theorems more easy to understand and to remember. *Algebraical* proofs of course introduce the difficulty of 'incommensurables.'

Min. We will now take the new Propositions, &c. which have been suggested.

Here is an Axiom :—'*Two lines cannot have a common segment.*'

Euc. Good. I have tacitly assumed it, but it may as well be stated.

Min. Several new Theorems have been suggested, but only two of them seem to me worth mentioning. They are :—

'*All right angles are equal.*'

Euc. I have already approved of that (p. 219).

Min. The other is one that is popular with most of your Rivals :—

'*Of all the Lines which can be drawn to a Line from a point without it, the perpendicular is least; and, of the rest, that which is nearer to the perpendicular is less than one more remote; and the lesser is nearer than the greater; and from the same point only two equal Lines can be drawn to the other Line, one on each side of the perpendicular.*'

Euc. I like it on the whole, though so long an enunciation will be alarming to beginners. But it is strictly analogous to III. 7. Introduce it by all means in the revised edition of my Manual. It will be well, however, to lay it down as a general rule, that no Proposition shall be so interpolated, unless it be of such importance and value as to be thought worthy of being quoted as proved, in the same way in which candidates in examinations are now allowed to quote Propositions of mine.

Min. (*with a fearful yawn*) Well! I have no more to say.

§ 7. *The summing-up.*

Euc. 'The cock doth craw, the day doth daw,' and all respectable ghosts ought to be going home. Let me carry with me the hope that I have convinced you of the importance, if not the necessity, of retaining my order and numbering, and my method of treating straight Lines, angles, right angles, and (most especially) Parallels. Leave me these untouched, and I shall look on with great contentment while other changes are made—while my proofs are abridged and improved—while alternative proofs are appended to mine—and while new Problems and Theorems are interpolated.

In all these matters my Manual is capable of almost unlimited improvement.

[*To the sound of slow music,* EUCLID *and the other ghosts 'heavily vanish,' according to Shakespeare's approved stage-direction.* MINOS *wakes with a start, and betakes himself to bed, 'a sadder and a wiser man.'*]

APPENDICES.

APPENDIX I.

Extract from Mr. Todhunter's essay on 'Elementary Geometry,' included in 'The Conflict of Studies, &c.'

It has been said by a distinguished philosopher that England is "usually the last to enter into the general movement of the European mind." The author of the remark probably meant to assert that a man or a system may have become famous on the continent, while we are almost ignorant of the name of the man and the claims of his system. Perhaps, however, a wider range might be given to the assertion. An exploded theory or a disadvantageous practice, like a rebel or a patriot in distress, seeks refuge on our shores to spend its last days in comfort if not in splendour. Just when those who originally set up an idol begin to suspect that they have been too extravagant in their devotions we receive the discredited image and commence our adorations. It is a less usual but more dangerous illustration of the principle, if just as foreigners are learning to admire one of our peculiarities we should grow weary of it.

In teaching elementary geometry in England we have for a long time been accustomed to use the well-known *Elements of Euclid*. At the present moment, when we learn from the best testimony, namely, the admission of anti-Euclideans, that both in France and Italy dissatisfaction is felt with the system hitherto used, accompanied with more or less desire to adopt ours, we are urged by many persons to exchange our system for one which is falling out of favour on the continent.

. . . .

Many assertions have been made in discussion which rest entirely on the authority of the individual advocate, and thus it is necessary to be somewhat critical in our estimate of the value of the testimony. Two witnesses who are put prominently forward are MM. Demogeot and Montucci, who drew up a report on English education for the French Government. Now I have no doubt that these gentlemen were suited in some respects to report on English education, as they were selected for that purpose; but I have searched in vain for any evidence of their special mathematical qualifications. No list of mathematical publications that I have consulted has ever presented either of these names, and I am totally at a loss to conceive on what grounds an extravagant respect has been claimed for their opinions. The following sentence has been quoted with approbation from these writers: "Le trait distinctif de l'enseignement des mathématiques en Angleterre c'est qu'on y fait appel plutôt à la mémoire qu'à l'intelligence de l'élève." In the first place we ought to know on what evidence this wide generalisation is constructed. Did the writers visit some of the humbler schools in England in which the elements of arithmetic and mensuration were rudely taught, and draw from this narrow experience an inference as to the range of mathematical instruction throughout England? Or did they find on inspecting some of our larger public schools that the mathematical condition was unsatisfactory? In the latter case this might have arisen from exclusive devotion to classics, or from preference for some of the fashionable novelties of the day, or from want of attention and patience in the teachers. On the most unfavourable supposition the condemnation pronounced on the general mathematical training in England cannot be justified. But take some kind of experimental test. Let an inquirer carefully collect the mathematical examination papers issued throughout England in a single year, including those proposed at the Universities and the Colleges, and those set at the Military Examinations, the Civil Service Examinations, and the so-called Local Examinations. I say then, without fear of contradiction, that the original problems and examples contained in these papers will for interest, variety, and ingenuity surpass any similar set that could be found in any country of the world. Then any person practically conversant with teaching and examining can judge whether the teaching is likely to be the worst where the examining is the most excellent.

The sentence quoted from MM. Demogeot and Montucci, in order to have any value, ought to have proceeded from writers more nearly on a level with the distinguished mathematical teachers in England. So far as any foundation can be assigned for this statement, it will probably apply not to mathematics especially but to all our studies, and amount to this, that our incessant examinations lead to an over cultivation of the memory. Then as to the practical bearing of the remark on our present subject it is obvious that the charge, if true, is quite independent of the text-book used for instruction, and might remain equally valid if Euclid were exchanged for any modern author.

The French gentlemen further on contrast what they call Euclid's verbiage with the elegant conciseness of the French methods. It is surely more than an answer to these writers to oppose the high opinion of the merits of Euclid expressed by mathematicians of European fame like Duhamel and Hoüel. See the *First Report of the Association for the Improvement of Geometrical Teaching*, p. 10.

When we compare the lustre of the mathematical reputation of these latter names with the obscurity of the two former, it seems that there is a great want of accuracy in the statement made in a recent circular: 'The opinion of French mathematicians on this question, is plainly expressed in the Report of MM. Demogeot and Montucci. . . .'

I should have to quote very largely indeed if I wished to draw attention to every hazardous statement which has been advanced; I must therefore severely restrain myself. Consider the following: 'Unquestionably the best teachers depart largely from his words, and even from his methods. That is, they use the work of Euclid, but they would teach better without it. And this is especially true of the application to problems. Everybody recollects, even if he have not the daily experience, how unavailable for problems a boy's knowledge of Euclid generally is.' The value of such a statement depends entirely on the range of the experience from which it has been derived. Suppose for instance that the writer had been for many years an examiner in a large University in which against each candidate's name the school was recorded from which he came; suppose that the writer had also been much engaged in the numerous examinations connected with the military institutions; suppose that he had also been for a quarter of a century in residence at one of the largest colleges at Cambridge, and actively employed in the tuition; suppose also that it had been his duty to classify the new students for lecture purposes by examining them in Euclid and other parts of elementary mathematics; and finally suppose that he was in constant communication with the teachers in many of the large schools: then his opinion would have enjoyed an authority which in the absence of these circumstances cannot be claimed for it.

If I may venture to refer to my own experience, which I fear commenced when the writer whom I have just quoted was in his cradle, I may say that I have taught geometry both Euclidean and non-Euclidean, that my own early studies and prepossessions were towards the latter, but that my testimony would now be entirely in favour of the former.

. . . .

I admit that to teach Euclid requires patience both from the tutor and the pupil; but I can affirm that I have known many teachers who have succeeded admirably, and have sent a large number of pupils to the University well skilled in solving deductions and examples; nor have I ever known a really able and zealous teacher to fail. I am happy to supplement my own testimony by an extract from the very interesting lecture on Geometrical Teaching by Dr. Lees, of St. Andrews. 'Whatever may be the cause of failure in England, it is clear as any demonstration can be that the failure cannot be ascribed to Euclid. Because in Scotland we do employ Euclid as the text-book for our students, and in Scotland we have the teaching of Geometry attended with the most complete success; and this not only in the colleges, but in all the higher and more important schools and academies of the country, and in many of the parish schools even, where the attention of the teacher is necessarily so much divided.' See also the remarkable *Narrative-Essay on a Liberal Education*, by the Rev. S. Hawtrey, A.M., Assistant-Master, Eton.

. . . .

During the existence of the East India Company's military college at Addiscombe, it is well known that the cadets were instructed in mathematics by the aid of a course drawn up by the late Professor Cape. The geometry in this course was of the kind which our modern reformers recommend, being founded on Legendre, and adopting the principle of hypothetical constructions which is now so emphatically praised. In certain large schools where youths were trained for the military colleges it was usual to instruct a class of candidates for Woolwich, in Euclid, and a class of candidates for Addiscombe in Cape's adaptation of Legendre. Fairness in the procedure was secured by giving the same number of hours by the same masters to each class; and the honour and rewards which attended success supplied an effectual stimulus both to teachers and pupils. Now consider the result. I was assured by a teacher who was for many years distinguished for the number and the success of his pupils, that the training acquired by the Euclid class was far superior to that acquired by the Legendre class. The Euclid was not more difficult to teach and was more potent and more beneficial in its influence. The testimony made the stronger impression on me because at the time I was disposed from theoretical considerations to hold an opposite opinion; I was inclined for example to support the use of hypothetical constructions. Such experience as I afterwards gained shewed the soundness of the judgment at which the practical teacher had arrived; and I have also received the emphatic evidence of others who had good opportunities of considering the question, and had come to the same conclusion. I have myself examined at Woolwich and at Addiscombe, and am confident that the teaching in both institutions was sound and zealous; but I have no hesitation in saying that the foundation obtained from Euclid was sounder than that from Legendre.

. . . .

Although I have admitted that the study of Euclid is one that really demands patient attention from the beginner, yet I cannot admit that the tax is unreasonable. My own experience has been gained in the following manner. Some years since on being appointed principal mathematical lecturer in my college, more systematic arrangements were introduced for the lectures of the freshmen than had been previously adopted; and as the Euclid seemed to be one of the less popular subjects I undertook it myself. Thus for a long period the way in which this has been taught in schools, and the results of such teaching, have been brought under my notice. It need scarcely be said that while many of the students who have thus presented themselves to me have been distinguished for mathematical taste and power, yet the majority have been of other kinds; namely, either persons of ability whose attention was fully occupied with studies different from mathematics, or persons of scanty attainments and feeble power who could do little more than pass the ordinary examination. I can distinctly affirm that the cases of hopeless failure in Euclid were very few; and the advantages derived from the study, even by men of feeble ability, were most decided.

In comparing the performance in Euclid with that in Arithmetic and Algebra there could be no doubt that the Euclid had made the deepest and most beneficial impression: in fact it might be asserted that this constituted by far the most valuable part of the whole training to which such persons were subjected. Even the modes of expression in Euclid, which have been theoretically condemned as long and wearisome, seemed to be in practice well adapted to the position of beginners. As I have already stated there appears to me a decided improvement gradually taking place in the knowledge of the subject exhibited by youths on entering the University. My deliberate judgment is that our ordinary students would suffer very considerably if instead of the well-reasoned system of Euclid any of the more popular but less rigid manuals were allowed to be taken as a substitute.

Let me now make a few remarks on the demand which has been made to allow other books instead of Euclid in *examinations*. It has been said: "We demand that we should not be,—as we are now, by the fact of Euclid being set as a text-book for so many examinations,—practically obliged to adhere to one book. Surely such a request, made by men who know what they want, and are competent to form an opinion on the subject,—and made in earnest,—should induce the Universities and other examining bodies to yield their consent. The grounds of the demand then are three; that it is made in earnest, that it is made by those who know what they want, and that it is made by those who are competent to form an opinion on the subject. I need not delay on two of the grounds; the experience of every day shews that claimants may know what they want, and be terribly in earnest in their solicitations, and yet it may be the duty of those to whom the appeal is made to resist it. Moreover it is obvious that the adoption of Euclid as a text book is prescribed by those who are equally in earnest and know what they recommend. In short if no institution is to be defended when it is attacked knowingly and earnestly, it is plain that no institution is safe.

I turn then to the other ground, namely that the demand is made by men who are competent to form an opinion on the subject. Now it is not for me to affect to speak in the name of the University of Cambridge; mine is the opinion of only a private unofficial resident. But I have little doubt that many persons here will maintain, without questioning the competence of the claimants to form an opinion, that we ourselves are still much more competent to form an opinion.

For it will not be denied that in all which relates to mathematical knowledge we have an aggregate of eminence which far surpasses what has yet been collected together to press the demand on the University. Moreover as inspectors and judges we occupy a central position as it were, and thus enjoy opportunities which do not fall to isolated teachers however eminent and experienced. The incessant demands made upon the University to furnish examiners for schools and for the local examinations keep us as a body practically familiar with the standard of excellence attained in various places of instruction. Then as college lecturers and private tutors we have the strongest motives for keenly discriminating the state of mathematical knowledge in different schools, as shewn by the performance of the candidates when brought under our notice. Moreover some of the residents in the University by continued intercourse with old pupils, now themselves occupying important positions as teachers, are enabled to prolong and enlarge the experience which they may have already obtained directly or indirectly. If it is obvious that certain teachers by ability and devotion have for many years sent up well-trained pupils, the University may well consider that it would be neither right nor wise to deprive its best friends of their justly earned distinction, by relaxing in any way the rigour of the examinations. Instead then of urging an instant acquiescence with demands on the ground that those who make them are well qualified to judge, the claimants should endeavour by *argument* to convince others who are still better qualified to judge.

Here let me invite attention to the following remark which has been made in support of the claim: "In every other subject this is freely accorded; we are not obliged to use certain grammars or dictionaries, or one fixed treatise on arithmetic, algebra, trigonometry, chemistry, or any other branch of science. Why are we to be tied to one book in geometry alone?" Now in the first place it may be said that there are great advantages in the general use of one common book; and that when one book has long been used almost exclusively it would be rash to throw away certain good in order to grasp at phantasmal benefits. So well is this principle established that we have seen in recent times a vigorous, and it would seem successful, effort to secure the use of a common Latin Grammar in the eminent public schools. In the second place the analogy which is adduced in the remark quoted above would be rejected by many persons as involving an obvious fallacy, namely that the word *geometry* denotes the same thing by all who use it. By the admirers of Euclid it means a system of demonstrated propositions valued more for the process of reasoning involved than for the results obtained. Whereas with some of the modern reformers the rigour of the method is of small account compared with the facts themselves. We have only to consult the modern books named in a certain list, beginning with the *Essentials of Geometry*, to see that practically the object of some of our reformers is not to teach the same subject with the aid of a different text-book, but to teach something very different from what is found in Euclid, under the common name of geometry.

It may be said that I am *assuming* the point in question, namely, that Euclid is the best book in geometry; but this is not the case. I am not an advocate for *finality* in this matter; though I do go so far as to say that a book should be *decidedly* better than Euclid before we give up the advantages of uniformity which it will be almost impossible to secure if the present system is abandoned. But, as it has been well observed by one of the most distinguished mathematicians in Cambridge,

"The demand is unreasonable to throw aside Euclid in favour of any compendium however meagre and however unsound; and this is really the demand which is made: it will be time enough to consider about the discontinuance of Euclid when a better book is deliberately offered." It may be added that the superiority to Euclid must be established by indisputable evidence; not by the author's own estimation, the natural but partial testimony of parental fondness; not by the hasty prediction of some anonymous and irresponsible reviewer; not by the authority of eminent men, unless the eminence is founded on mathematical attainments; not even by the verdict of teachers who are not conspicuous for the success of their pupils. The decision must rest with students, teachers, and examiners, of considerable reputation in the range of the mathematical sciences.

It must be allowed that there is diversity of opinion among the opponents of Euclid, for while the majority seem to claim freedom in the use of any text-books they please, others rather advocate the construction and general adoption of a new text-book. The former class on the whole seem to want something easier and more popular than Euclid; among the latter class there are a few whose notion seems to be that the text-book should be more rigorous and more extensive than Euclid. There are various considerations which seem to me to indicate that if a change be made it will not be in the direction of *greater rigour*; the origin of the movement, the character of the text-books which have hitherto been issued, and the pressure of more modern and more attractive studies, combine to warn us that if the traditional authority which belongs to Euclid be abandoned geometry will be compelled to occupy a position in general education much inferior to that which it now holds.

. . . .

There is one very obvious mode of advancing the cause of the anti-Euclidean party, which I believe will do far more for them than the most confident assertions and predictions of the merits of the course which they advocate: let them train youths on their system to gain the highest places in the Cambridge Mathematical Tripos, and then other teachers will readily follow in the path thus opened to distinction. But it may naturally be said that as long as Euclid is prescribed for the text-book, the conditions of competition are unfair towards those who adopt some modern substitute; I will examine this point. In the Cambridge Examination for Mathematical Honours there are at present sixteen papers; a quarter of the first paper is devoted to book-work questions on Euclid. Now suppose that 1000 marks are assigned to the whole examination, and that about five of these fall to the book-work in Euclid. A student of any modern system would surely be able to secure some of these five marks, even from a stern Euclidean partisan. But to take the worst case, suppose the candidate deliberately rejects all chance of these five marks, and turns to the other matter on the paper, especially to the problems; here the advantage will be irresistibly on his side owing to the "superiority of the modern to the ancient methods of geometry" which is confidently asserted. It must be remembered that in spite of all warning and commands to the contrary, examiners will persist in making their papers longer than can be treated fully in the assigned time, so that the sacrifice of the book-work will be in itself trifling and will be abundantly compensated by the greater facility at the solution of problems which is claimed for the modern teaching, as compared with the "unsuggestiveness" of Euclid, and by the greater accuracy of reasoning, since we are told that "the logical training to be got from Euclid is very imperfect and in some respects bad." Thus on the whole the disciple of the modern school will even in the first paper of the Cambridge Tripos Examination be more favourably situated than the student of Euclid; and of course in the other papers the advantages in his favour become largely increased. For we must remember that we are expressly told that Euclid is "an unsuitable preparation for the higher mathematical training of the present day;" and that "those who continue their mathematical reading with a view of obtaining honours at the University . . . will gain much through economy of time and the advantage of modern lights."

The final result is this; according to the promises of the geometrical reformers, one of their pupils might sacrifice five marks out of a thousand, while for all the remaining 995 his chance would be superior to that of a Euclid-trained student. It may be added that in future the Cambridge Mathematical Examinations are to be rather longer than they have been up to the date of my writing; so that the advantage of the anti-Euclidean school will be increased. Moreover we must remember that in the Smith's Prizes Examination the elementary geometry of Euclid scarcely appears, so that the modern reformers would not have here any obstacle to the triumphant vindication of their superiority as teachers of the higher mathematics. The marvellous thing is that in these days of competition for educational prizes those who believe themselves to possess such a vast superiority of methods do not keep the secret to themselves, instead of offering it to all, and pressing it on the reluctant and incredulous. Surely instead of mere *assertion* of the benefits to be secured by the modern treatment, it will be far more dignified and far more conclusive to *demonstrate* the proposition by brilliant success in the Cambridge Mathematical Tripos. Suppose we were to read in the ordinary channels of information some such notice as this next January: "The first six wranglers are considered to owe much of their success to the fact that in their training the fossil geometry of Alexandria was thrown aside and recent specimens substituted;" then opposition would be vanquished, and teachers would wonder, praise, and imitate. But until the promises of success are followed by a performance as yet never witnessed we are reminded of the case of a bald hairdresser who presses on his customers his infallible specific for producing redundant locks.

. . . .

To those who object to Euclid as an inadequate course of plane geometry it may then be replied briefly that it is easy, if thought convenient or necessary, to supply any additional matter. But for my part I think there are grave objections to any large increase in the extent of the course of synthetical geometry which is to be prepared for examination purposes. One great drawback to our present system of mathematical instruction and examination is the monotony which prevails in many parts. When a mathematical subject has been studied so far as to master the essential principles, little more is gained by pursuing these principles into almost endless applications. On this account we may be disposed to regard with slender satisfaction the expenditure of much time on geometrical conic sections; the student seems to gain only new facts, but no fresh ideas or principles. Thus after a moderate course of synthetical geometry such as Euclid supplies, it may be most advantageous for the student to pass on to other subjects like analytical geometry and trigonometry which present him with ideas of another kind, and not mere repetitions of those with which he is already familiar.

. . . .

It has been said, and apparently with great justice, that examination in elementry geometry under a system of unrestricted text-books will be a very troublesome process; for it is obvious that in different systems the demonstration of a particular proposition may be more or less laborious, and so may be entitled to more or fewer marks. This perplexity is certainly felt by examiners as regards geometrical conic sections; and by teachers also who may be uncertain as to the particular system which the examiners may prefer or favour. It has been *asserted* that the objection thus raised is imaginary, and that "the manuals of geometry will not differ from one another nearly so widely as the manuals of algebra or chemistry: yet it is not difficult to examine in algebra and chemistry." But I am unable to feel the confidence thus expressed. It seems to me that much more variety may be expected in treatises on geometry than on algebra; certainly if we may judge from the experience of the examiners at Cambridge the subject of geometrical conics is the most embarrassing which occurs at present, and this fact suggests a conclusion very different from that which is laid down in the preceding quotation. Of course there will be no trouble in examining a single school, because the system there adopted will be known and followed by the examiner.

I have no wish to exaggerate the difficulty; but I consider it to be real and serious, more especially as it presents itself at the outset of a youth's career, and so may cause disappointment just when discriminating encouragement is most valuable. But I think the matter must be left almost entirely to the discretion of examiners; the attempts which have been made to settle it by regulation do not seem to me very happy. For example, I read: "As the existing text-books are not very numerous, it would not be too much to require examiners to be acquainted with them sufficiently for the purpose of testing the accuracy of written, or even, if necessary, of oral answers." The language seems to me truly extraordinary. Surely examiners are in general men of more mathematical attainments than this implies; for it would appear that all we can expect them to do is to turn to some text-book and see if the student has correctly reproduced it. The process in a *viva voce* examination would be rather ignominious if when an answer had been returned by a candidate some indifferent manual had to be consulted to see if the answer was correct.

I have heard that an examining board has recently issued instructions to its executive officers to make themselves acquainted with the various text-books. This does not enjoin distinctly, what the above quotation implies, that the examiner is to accept all demonstrations which are in print as of nearly equal value; but it seems rather to suggest such a course. The point is important and should be settled. Suppose a candidate offered something taken from the *Essentials of Geometry*, and the examiner was convinced that the treatment was inadequate or unsound; then is the candidate nevertheless to obtain full marks? Again, it may be asked, why printed books alone are to be accepted; and why a student who has gone through a manuscript course of geometry should be precluded from following it? The regulation might be made that he should submit a copy of his manuscript course to the examiner in order that it might be ascertained whether he had reproduced it accurately. As I have already intimated, the only plan which can be adopted is to choose able and impartial men for examiners, and trust them to appreciate the merits of the papers submitted by the candidate to them.

The examiners will find many perplexing cases I have no doubt; one great source of trouble seems to me to consist in the fact that what may be a sound demonstration to one person with adequate preliminary study is not a demonstration to another person who has not gone through the discipline. To take a very simple example: let the proposition be, *The angles at the base of an isosceles triangle are equal.* Suppose a candidate dismisses this briefly with the words, *this is evident from symmetry*; the question will be, what amount of credit is to be assigned to him. It is quite possible that a well-trained mathematician may hold himself convinced of the truth of the proposition by the consideration of symmetry, but it does not follow that the statement would really be a demonstration for an early student. Or suppose that another imbued with "the doctrine of the imaginary and inconceivable" says as briefly "the proposition is true, for the inequality of the angles is inconceivable and therefore false;" then is the examiner to award full marks, even if he himself belongs to the school of metaphysics which denies that the inconceivable is necessarily the false?

. . . .

It has been urged as an objection against Euclid that the number of his propositions is too great. Thus it has been said that the 173 propositions of the six books might be reduced to 120, and taught in very little more than half the time required to go through the same matter in Euclid. So far as the *half time* is concerned this seems to be only an expression of belief as to the result of an untried experiment; it is based on the comparison of a few other books with Euclid, one of these being the Course of the late Professor Cape; as I have already stated, actual experience suggests a conclusion directly contrary to the present prediction. As to the *number* of propositions we readily admit that a reduction might be made, for it is obvious that we may in many cases either combine or separate according to our taste. But the difficulty of a subject does not vary directly as the number of propositions in which it is contained; a single proposition will in some cases require more time and attention than half a dozen others. I have no doubt that the mixture of easy propositions with the more difficult is a great encouragement to beginners in Euclid; and instead of diminishing the number of propositions I should prefer to see some increase: for example I should like to have Euclid i. 26 divided into two parts, and Euclid i. 28 into two parts.

Again, it has been said that Euclid is artificial, and that he "has sacrificed to a great extent simplicity and naturalness in his demonstrations;" it is a curious instance of the difference of opinion which we may find on the same subject, for, with a much wider experience than the writer whom I quote, I believe that Euclid maintains, and does not sacrifice, simplicity and naturalness in his system, assuming that we wish to have strictness above all things.

The exclusion of hypothetical constructions has been represented as a great defect in Euclid; and it has been said that this has made the *confused order* of his first book necessary. Confused order is rather a contradictory expression; but it may be presumed that the charge is intended to be one of confusion: I venture to deny the confusion. I admit that Euclid wished to make the subject depend on as few axioms and postulates as possible; and this I regard as one of his great merits; it has been shown by one of the most distinguished mathematicians of our time how the history of science teaches in the clearest language that the struggle against self-imposed restrictions has been of the most signal service in the advancement of knowledge.

The use of hypothetical constructions will not present itself often enough to produce any very great saving in the demonstrations; while the difficulty which they produce to many beginners, as shown by the experience to which I have already referred, is a fatal objection to them. Why should a beginner not assume that he can draw a circle through four given points if he finds it convenient?

. . . .

Finally, I hold that Euclid, in his solution of the problems he requires, supplies matter which is simple and attractive to beginners, and which therefore adds practically nothing to their labours, while it has the advantage of rendering his treatise far more rigorous and convincing to them.

The objections against Euclid's order seem to me to spring mainly from an intrusion of natural history into the region of mathematics; I am not the first to print this remark, though it occurred to me independently. It is to the influence of the classificatory sciences that we probably owe this notion that it is desirable or essential in our geometrical course to have all the properties of triangles thrown together, then all the properties of rectangles, then perhaps all the properties of circles; and so on. Let me quote authority in favour of Euclid, far more impressive than any which on this point has been brought against him: "Euclid . . . fortunately for us, never dreamed of a geometry of triangles as distinguished from a geometry of circles, . . . but made one help out the other as he best could."

Euclid has been blamed for his adherence to the syllogistic method; but it is not necessary to say much on this point, because the reformers are not agreed concerning it: those who are against the syllogism may pair off with those who are for the syllogism. We are told in this connexion that, "the result is, as every one knows, that boys may have worked at Euclid for years, and may yet know next to nothing of Geometry." We may readily admit that such may be the case with boys exceptionally stupid or indolent; but if any teacher records this as the average result of his experience, it must I think be singularly to his own discredit.

There is, I see, a notion that the syllogistic form, since it makes the demonstrations a little longer, makes them more difficult; this I cannot admit. The number of words employed is not a test of the difficulty of a demonstration. An examiner, especially if he is examining *viva voce*, can readily find out where the difficulties of the demonstrations really lie; my own experience leads me to the conclusion that the syllogistic form instead of being an impediment is really a great assistance, especially to early students.

"Unsuggestiveness" has been urged as a fault in Euclid; which is interpreted to mean that it does not produce ability to solve problems. We are told: "Everybody recollects, even if he have not the daily experience, how unavailable for problems a boy's knowledge of Euclid generally is. Yet this is the true test of geometrical knowledge; and problems and original work ought to occupy a much larger share of a boy's time than they do at present." I need not repeat what I have already said, that English mathematicians, hitherto trained in Euclid, are unrivalled for their ingenuity and fertility in the construction and solution of problems. But I will remark that in the important mathematical examinations which are conducted at Cambridge the rapid and correct solution of problems is of paramount value, so that any teacher who can develop that power in his pupils will need no other evidence of the merits of his system.

Euclid's treatment of proportion has been especially marked out for condemnation; indeed, with the boldness which attaches to many assertions on the subject of elementary geometry, it has been pronounced already *dead*. In my own college it has never been laid aside; only a few months since one of our most influential tutors stated that he was accustomed to give a proposition out of the fifth book of Euclid to some candidates for emoluments, and he considered it a very satisfactory constituent of the whole process of testing them.

I should exceedingly regret the omission of the fifth book of Euclid, which I hold to be one of the most important parts of the training supplied by Elementary Geometry. I do not consider it necessary for beginners to go through the entire book; but the leading propositions might be mastered, and the student led to see how they can be developed if necessary. I may refer here to some valuable remarks which have been made on the subject by the writer of a *Syllabus of Elementary Geometry* . . . who himself I believe counts with the reformers. He sums up thus ". . . any easy and unsatisfactory short cuts (and I have sometimes seen an inclination for such) should be scouted, as a simple deception of inexperienced students."

However, I must remark that I see with great satisfaction the following *Resolution* which was adopted at a recent meeting of the *Association for the Improvement of Geometrical Teaching*: "That no treatise on Geometry can be regarded as complete without a rigorous treatment of ratio and proportion either by Euclid's method, or by some equally rigorous method of limits." It would be injudicious to lay much stress on resolutions carried by a majority of votes; but at least we have a striking contradiction to the confident statement that Euclid's theory of proportion is *dead*. We shall very likely see here, what has been noticed before, that a course may be proposed which differs widely from Euclid's, and then, under the guidance of superior knowledge and experience, the wanderers are brought back to the old path. Legendre's return to Euclid's treatment of parallels is a conspicuous example; see the valuable paper by Professor Kelland on *Superposition* in the *Edinburgh Transactions*, Vol. XXI.

I cannot avoid noticing one objection which has been urged

against Euclid in relation to his doctrine of proportion ; namely, that it leaves "the half-defined impression that all profound reasoning is something far fetched and artificial, and differing altogether from good clear common sense." It appears to me that if a person imagines that "good clear common sense" will be sufficient for mastering pure and mixed mathematics, to say nothing of contributing to their progress,—the sooner he is undeceived the better. Mathematical science consists of a rich collection of investigations accumulated by the incessant labour of many years, by which results are obtained far beyond the range of unassisted common sense; of these processes Euclid's theory of proportion is a good type, and it may well be said that from the degree of reverent attention which the student devotes to it, we may in most cases form a safe estimate of his future progress in these important subjects.

. . . .

In conclusion I will say that no person can be a warmer advocate than I am for the *improvement of Geometrical Teaching*: but I think that this may be obtained without the hazardous experiment of rejecting methods, the efficacy of which a long experience has abundantly demonstrated.

APPENDIX II.

Extract from Mr. De Morgan's review of Mr. Wilson's Geometry, in the 'Athenæum' for July 18, 1868.

The Schools' Inquiry Commission has raised the question whether Euclid be, as many suppose, the best elementary treatise on geometry, or whether it be a mockery, delusion, snare, hindrance, pitfall, shoal, shallow, and snake in the grass.

. . . .

We pass on to a slight examination of Mr. Wilson's book. We specially intend to separate the logician from the geometer. In the author's own interest, and that he may be as powerful a defender as can be of a cause we expect and desire to see fully argued, we recommend him to revise his notions of logic. We know that mathematicians care no more for logic than logicians for mathematics. The two eyes of exact science are mathematics and logic: the mathematical sect puts out the logical eye, the logical sect puts out the mathematical eye; each believing that it sees better with one eye than with two. The consequences are ludicrous. On the one side we have, by confusion of words, the great logician Hamilton bringing forward two quantities which are 'one and the same quantity,' Breadth and Depth, while, within a few sentences, 'the greater the Breadth, the less the Depth.' On the other side, we have the great mathematician, Mr. Wilson, also by confusion of words, speaking of the 'invariably syllogistic form of his [Euclid's] reasoning,' and, to show that this is not a mere slip, he afterwards talks of the 'detailed syllogistic form' as a 'source of obscurity to beginners, and damaging to true geometrical freedom and power.'

Euclid a book of syllogistic form! We stared. We never heard of such a book, except the edition of Herlinus and Dasypodius (1566), who, quite ignorant that Euclid was syllogistic already, made him so, and reckoned up the syllogisms. Thus i. 47 has 'syllogismi novem' at the head. They did not get much thanks; the book was never reprinted, and was in oblivion-dust when Hamilton mentioned the zealous but thick-headed logicians, as he called them. Prof. Mansel, in our own day, has reprinted one of their propositions as a curiosity. In 1831, Mr. De Morgan, advocating the reduction of a few propositions to detailed syllogistic form as an exercise for students, gave i. 47 as a specimen, in the Library of Useful Knowledge; this was reprinted, we believe, in the preface to various editions of Lardner's Euclid. A look will show the difference between Euclid and syllogistic form. Had the Elements been syllogistic, it would have been quoted in all time, as a proof of the rapid diffusion of Aristotle's writings, that they had saturated his junior contemporary with their methods: with a controversy, most likely, raised by those who would have contended that Euclid invented syllogistic form for himself. Now it is well known that diffusion of Aristotle's writings commenced after his death, and that it was—not quite correctly—the common belief that evulgation did not take place until two hundred years after his death. Could this belief ever have existed if Euclid had invariably used 'syllogistic form'?

What could have been meant? Craving pardon if wrong, we suspect Mr. Wilson to mean that Euclid did not deal in arguments with *suppressed premisses.* Euclid was quite right: the first reasonings presented to a beginner should be of full statement. He may be trained to suppression: but the true way to abbreviation is from the full length. Mr. Wilson does not use the phrases of reasoning consistently. He tells the student that a *corollary* is 'a geometrical truth easily deducible from a theorem': and then, to the theorem that only one perpendicular can be drawn to a straight line, he gives as a corollary that the external angle of a triangle is greater than the internal opposite. This is not a corollary from the theorem, but a matter taken for granted in proving it.

Leaving this, with a recommendation to the author to strengthen his armour by the study of logic, we pass on to the system. There is in it one great point, which brings down all the rest if it fall, and may perhaps—but we must see Part II. before we decide—support the rest if it stand. That point is the treatment of the angle, which amounts to this, that certain notions about *direction*, taken as self-evident, are permitted to make all about angles, parallels and all, immediate consequences. The notion of continuous change, and consequences derived from it, enters without even an express assumption: 'continually' is enough.

Mr. Wilson would not have ventured expressly to postulate that when a magnitude changes continuously, all magnitudes which change with it also change continuously. He knows that when a point moves on a line, an angle may undergo a sudden change of two right angles. He trusts to the beginner's perception of truth in the case before him: the whole truth would make that beginner feel that he is on a foundation of general principles made safe for him by selection, and only safe because the exceptions are not likely to occur to his mind. On this we write, as Newton wrote on another matter, *Falsa! Falsa! Non est Geometria!*

What 'direction' is we are not told, except that 'straight lines which meet have different directions.' Is a direction a magnitude? Is one direction greater than another? We should suppose so; for an angle, a magnitude, a thing which is to be halved and quartered, is the 'difference of the direction' of 'two straight lines that meet one another.' A better definition follows; the 'quantity of turning' by which we pass from one direction to another. But hardly any use is made of this, and none at the commencement. And why two definitions? Is the difference of two directions the *same thing* as the rotation by which we pass from one to the other? Is the difference of position of London and Rugby a number of miles on the railroad? Yes, in a loosely-derived and popular and slip-slop sense: and in like manner we say that one man *is* a pigeon-pie, and another *is* a shoulder of lamb, when we describe their contributions to a pic-nic. But *non est geometria!* Metaphor and paronomasia can draw the car of poetry; but they tumble the waggon of geometry into the ditch.

Parallels, of course, are lines which have the same direction. It is stated, as an immediate consequence, that two lines which meet cannot make the same angle with a third line, on the same side, for they are in different directions. Parallels are knocked over in a trice. There is a covert notion of direction which, though only defined with reference to lines which meet, is straightway transferred to lines which do not. According to the definition, direction is a relation of lines which *do* meet, and yet lines which have the same direction can be lines which *never* meet. There is a great *quantity of turning* wanted; turning of implied assumption into expressed. Mr. Wilson would, we have no doubt, immediately introduce and defend all we ask for; and we quite admit that his system has a right to it. How do you know, we ask, that lines which have the same direction never meet? Answer—lines which meet have *different* directions. We know they have; but how do we know that, under the definition given, the relation called *direction* has any application at all to lines which never meet? The use of the notion of limits may give an answer: but what is the system of geometry which introduces continuity and limits to the mind as yet untaught to think of space and of magnitude? Answer, a royal road. If the difficulty were met by expressed postulates, the very beginner himself would be frightened.

There is a possibility that Mr. Wilson may mean that lines which make the same angle with a third on the same side are in the same direction. If this be the case, either he assumes that lines equally inclined to one straight line are equally inclined to all,—and this we believe he does, under a play on the word 'direction'; or he makes a quibble only one degree above a pun on his own arbitrary assumption of his right to the word 'same': and this we do not believe he does. He should have been more explicit: he should have said, My system involves an assumption which has lain at the root of many attempts upon the question of parallels, and has always been scouted as soon as seen. He should have added, I assume Euclid's eleventh axiom: I have a notion of direction; I tell you that lines which meet have different directions; I imply that lines which make different angles with a third have also different directions; and I assume that lines of different directions will meet. Mr. Wilson is so concise that it is not easy to be very positive as to how much he will admit of the above, or how he will get over or round it. When put upon his defence he must be more explicit. Mr. Wilson gives four explicit axioms about the straight line: and not one about the angle.

. . . .

We feel confidence that no such system as Mr. Wilson has put forward will replace Euclid in this country. The old geometry is a very English subject, and the heretics of this orthodoxy are the extreme of heretics: even Bishop Colenso has written a Euclid. And the reason is of the same kind as that by which the classics have held their ground in education. There is a mixture of good sense and of what, for want of a better name, people call prejudice: but to this mixture we owe our stability. The proper word is *postjudice*, a clinging to past experience, often longer than is held judicious by after times. We only desire to avail ourselves of this feeling until the book is produced which is to supplant Euclid; we regret the manner in which it has allowed the retention of the faults of Euclid; and we trust the fight against it will rage until it ends in an amended form of Euclid.

APPENDIX III.

Proof that, if any one Proposition of Table II be granted as an Axiom, the rest can be deduced from it. (See pp. 34, 40.)

"... and so we make it quite a merry-go-rounder." I was obliged to consider a little before I understood what Mr. Peggotty meant by this figure, expressive of a complete circle of intelligence.

It is to be proved that, if any *one* of the Propositions of Table II be granted, the rest can be proved.

It is assumed that the lesser of two unequal finite magnitudes of the same kind may be multiplied so as to exceed the greater.

Euclid I, 1 to 28, is assumed as proved.

It is assumed that, where two Propositions are Contranominals, so that each can be proved from the other, it is not necessary to include *both* in the series of proofs.

Lemma 1.

A Pair of Lines, of which one contains two points equidistant from the other, have a common perpendicular.

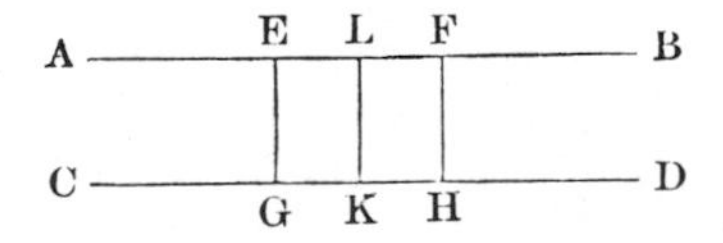

Let AB contain 2 points E, F, equidistant from CD. From E, F, draw EG, FH, $\perp$ CD; bisect GH in K, and EF in L, and join KL.

Now $EG = FH$; [*hyp.*

hence, if the diagram be reversed, and so placed on its former traces that G coincides with H, and H with G, K retaining its position, GE coincides with HF, and HF with GE;

$\therefore$ E coincides with F, and F with E;

$\therefore$ L retains its position;

$\therefore$ $\angle GKL$ coincides with $\angle HKL$, and is equal to it;

$\therefore$ $\angle$s at K are right.

Similarly $\angle$s at L are right.

Therefore a Pair of Lines, &c. Q. E. D.

(α). II. 1.

A Pair of separational Lines are equally inclined to any transversal.

[N.B. The Contranominal of this will be proved at the end of the series.]

(β). II. 16 (a).

Two intersecting Lines cannot both be separational from the same Line.

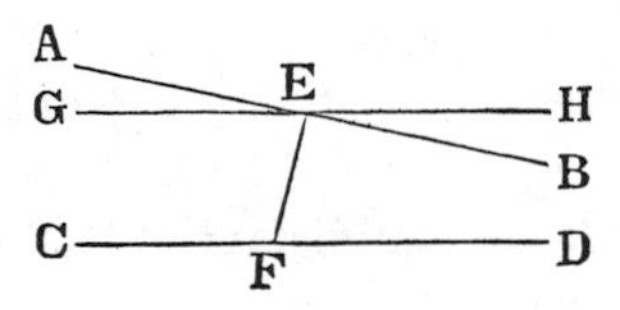

Let AEB, GEH be two intersecting Lines, and CD another Line. It is to be proved that they cannot both be separational from CD.

In CD take any point F; and join EF.

Now, if possible, let AB, GH both be separational from CD;

$\therefore$ $\angle$s AEF, GEF are both equal to $\angle EFD$; [(α).

$\therefore$ they are equal to each other; which is absurd.

Therefore two intersecting Lines &c. Q. E. D.

(γ). II. 6.

A Pair of separational Lines are equidistantial from each other.

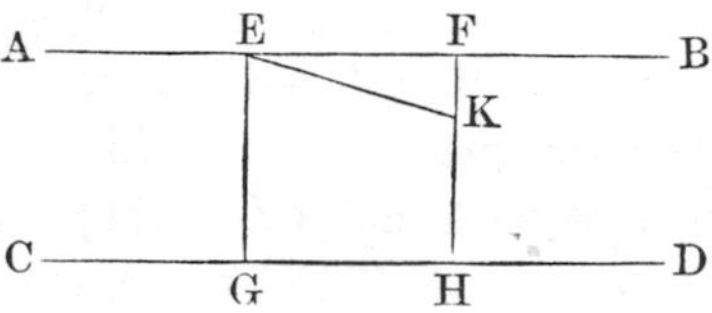

Let AB, CD be separational Lines: it shall be proved that they are equidistantial from each other.

In AB take any 2 points E, F; and draw EG, FH, $\perp CD$.

Now, if $FH > EG$, from it cut off KH equal to EG; and join EK;

then, $\because EG = KH$,

$\therefore EK$, CD have a common perpendicular; [Lemma 1.

$\therefore EK$ is separational from CD; [Euc. I. 27.

$\therefore AB$, EK, intersecting Lines, are both separational from CD; which is absurd; [(β).

$\therefore FH$ is not $> EG$.

Similarly it may be proved that EG is not $> FH$.

Therefore $EG = FH$.

Similarly it may be proved that any 2 points in AB are equidistant from CD, and that any 2 points in CD are equidistant from AB.

Therefore AB, CD are equidistantial from each other.

Therefore a Pair &c. Q. E. D.

(δ). II. 11.

A Pair of Lines, which are equally inclined to a certain transversal, are equidistantial from each other.

A Pair of Lines, which are equally inclined to a certain transversal, are separational; [Euc. I. 27.

also a Pair of separational Lines are equidistantial from each other; [(γ).

$\therefore$ a Pair of Lines, &c. Q. E. D.

(ε). II. 8.

Through a given point, without a given Line, a Line may be drawn such that the two Lines are equidistantial from each other.

For, if through the given point there be drawn a transversal, there can also be drawn through it a Line such that the two Lines make equal ∠s with the transversal; [Euc. I. 23.
and this Line will be such that the two Lines are equidistantial from each other. [(δ).
Therefore, &c. Q. E. D.

(ζ). II. 17.

A Line cannot recede from and then approach another; nor can one approach and then recede from another on the same side of it.

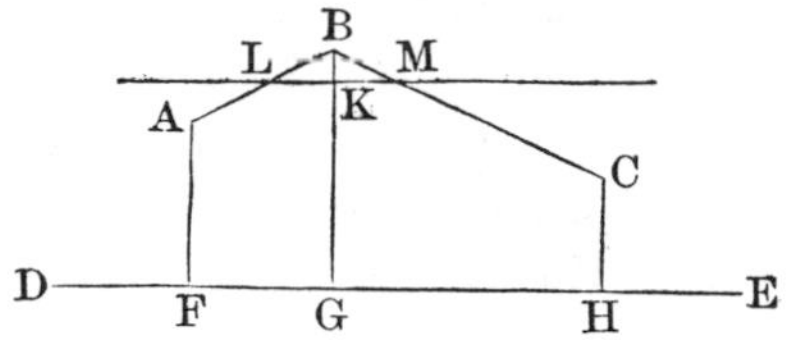

If possible, let *ABC* first recede from, and then approach, *DE*; that is, let the perpendicular *BG* be > each of the two perpendiculars *AF*, *CH*.

From *GB* cut off *GK* > each of the two, *AF*, *CH*.

Now a Line may be drawn, through *K*, equidistantial from *DE*; [(ε).
and the points *A*, *C* will lie on the side of it next to *DE*, and *B* on the other side;
∴ it will cut *AB* between *A* and *B*, and *BC* between *B* and *C*.

Let *L*, *M* be the points of intersection; and join *LM*;
∴ the 2 Lines *LBM*, *LKM* contain a space; which is absurd.

Similarly it may be proved that *ABC* cannot first approach and then recede from *DE* on the same side of it.

Therefore a Line &c. Q. E. D.

(η). II. 13.

A Pair of Lines, of which one has two points on the same side of, and equidistant from, the other, are equidistantial from each other.

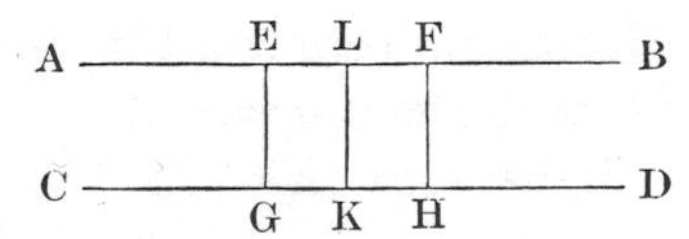

Let AB contain two points E, F, equidistant from CD. From E, F, draw EG, FH, $\perp CD$; bisect GH in K, and EF in L, and join KL.

Now $EG = FH$; [*hyp.*

hence, if the diagram be reversed, and so placed on its former traces that G coincides with H, and H with G, K retaining its position, GE coincides with HF, and HF with GE;

$\therefore$ E coincides with F, and F with E;

$\therefore$ L retains its position;

$\therefore$, if there be a point in LA whose distance is $< LK$, there is another such point in LB, and the Line AB will first recede from and then approach CD; which is absurd. [(ζ).

Similarly if there be one whose distance is $> LK$.

$\therefore$ AB is equidistantial from CD.

Similarly it may be proved that CD is equidistantial from AB.

Therefore a Pair of Lines, &c. Q. E. D.

Lemma 2.

Through a given point may be drawn a common perpendicular to a given Pair of Lines, of which each is equidistantial from the other.

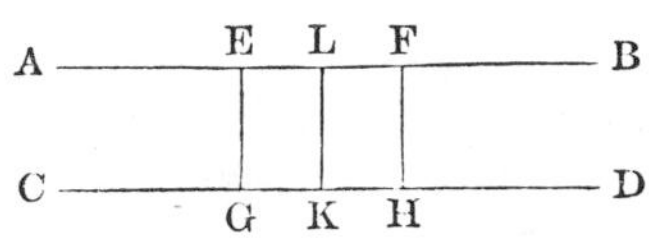

Let AB, CD be the given Pair of Lines.

Through the given point draw a Line perpendicular to AB, and let it meet AB in L. In AB take any 2 points E, F, equidistant from L. From E, F, draw EG, FH, perpendicular to CD. Bisect GH at K; and join KL.

Now E, F are 2 points, in AB, equidistant from CD; and GH is bisected in K, and EF in L;

$\therefore$ KL is a common perpendicular; [Lemma 1.

$\therefore$ it coincides with the Line drawn, through the given point, perpendicular to AB, since both meet AB at L;

$\therefore$ KL is the Line required.

Q. E. F.

(θ). II. 9.

A Pair of Lines, of which one has two points on the same side of, and equidistant from, the other, are equally inclined to any transversal.

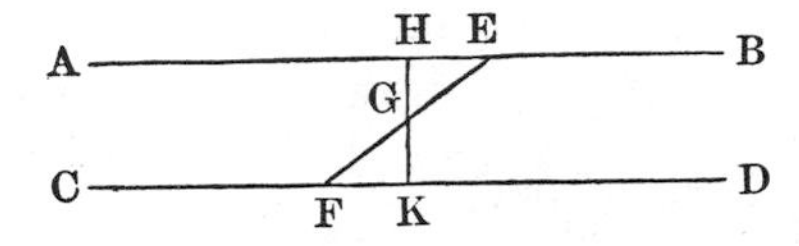

Let AB contain two points equidistant from CD, and let EF be a certain transversal: it shall be proved that $\angle AEF = \angle EFD$.

Now AB, CD, are equidistantial from each other. [(η).

Bisect EF at G; through G let HGK be drawn a common perpendicular to AB and CD. [Lemma 2.

Hence, in Triangles GEH, GFK, side GE and $\angle$s EGH, GHE, are respectively equal to side GF and $\angle$s FGK, GKF;

$\therefore$ $\angle GEH = \angle GFK$. [Euc. I. 26.

Therefore a Pair of Lines, &c. Q. E. D.

(κ). II. 3.

Through a given point, without a given Line, a Line may be drawn such that the two Lines are equally inclined to any transversal.

Take a second point, on the same side of the given Line and at the same distance from it; and join the 2 points.

Then the Line, so drawn, and the given Line, are equally inclined to any transversal. [(θ).

Therefore through a given point, &c. Q. E. D.

(λ). II. 18 (*b*).

The angles of a Triangle are together equal to two right angles.

Let ABC be a Triangle. It is to be proved that its 3 angles are together equal to 2 right angles.

Through A let DAE be drawn, such that DAE, BC are equally inclined to any transversal. [(κ).

Then $\angle B = \angle DAB$, and $\angle C = \angle EAC$;

$\therefore \angle s\ B, C, BAC = \angle s\ DAB, EAC, BAC$;

$= 2$ rt $\angle$ s. [Euc. I. 13.

Therefore the angles &c. Q. E. D.

(μ). II. 4.

A Pair of Lines, which are equally inclined to a certain transversal, are so to any transversal.

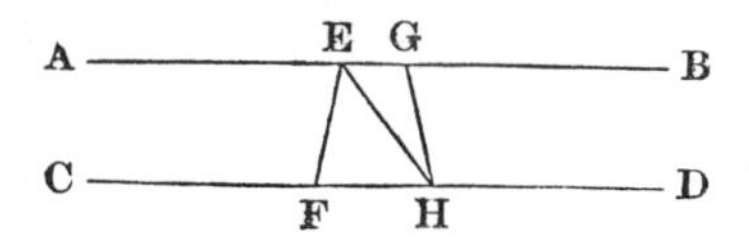

Let AB, CD be equally inclined to EF; and let GH be any other transversal. It shall be proved that they are equally inclined to GH.

Join EH.

Because $\angle$s of Triangle EFH together $= 2$ rt $\angle$s, and likewise those of Triangle EGH, [(λ).

$\therefore$ angles of Figure FG together $= 4$ rt angles;

also, by hypothesis, $\angle$s GEF, EFH together $= 2$ rt $\angle$s;

$\therefore$ remaining $\angle$s EGH, GHF together $= 2$ rt $\angle$s;

$\therefore$ AB, CD are equally inclined to GH.

Therefore a Pair of Lines, &c. Q. E. D.

CONTRANOMINAL OF (a). II. 2.

A Pair of Lines, which make with a third Line two interior angles, on one side of it, together less than two right angles, will meet on that side if produced.

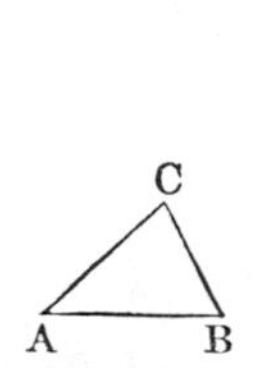

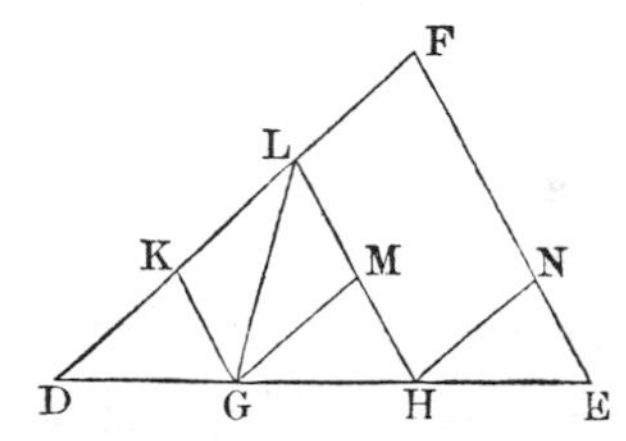

Let ABC, DEF be two Triangles such that $\angle$s, A, D are equal, and DE, DF equimultiples of AB, AC.

From DE cut off successive parts equal to AB; and let the points of section be G, H. At G, H make $\angle$s equal to $\angle E$.

Then the Lines, so drawn, are separational from EF and from one another; [Euc. I. 28.

$\therefore$ these Lines meet DF between D and F; call these points K, L.

At G, H make $\angle$s equal to $\angle D$.

Then the Lines, so drawn, are separational from DF;

$\therefore$ they respectively meet HL between H and L, and EF between E and F; call these points M, N.

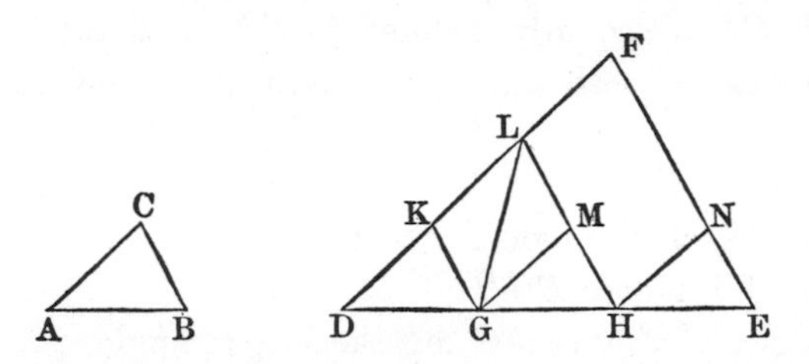

Because Triangles DGK, GHM, HEN are on equal bases and have their base-$\angle$s respectively equal,

$\therefore$ DK, GM, HN are equal. [Euc. I. 26.

Join GL.

Because DL, GM are equally inclined to DE,

$\therefore$ they are equally inclined to GL; [(μ).

$\therefore$ $\angle$s KLG, LGM are equal.

Similarly, $\because$ GK, HL are equally inclined to DE,

$\therefore$ they are equally inclined to GL;

$\therefore$ $\angle$s KGL, GLM are equal.

Because Triangles LGK, GLM are on same base LG and have their base-$\angle$s respectively equal,

$\therefore$ $KL = GM$, i.e. $= DK$. [I. 26.

Similarly it may be proved that $LF = HN$, i.e. $= DK$.

Hence DE, DF are equimultiples of DG, DK, i.e. of AB, DK;

but they are also equimultiples of AB, AC;

$\therefore$ $DK = AC$.

Because Triangles ABC, DGK have $\angle$s A, D equal, and AB, AC respectively equal to DG, DK,

$\therefore$ $\angle$s, B, DGK are equal, and likewise $\angle$s C, DKG. [I. 4.

Because GK, EF are equally inclined to DE,

$\therefore$ they are equally inclined to DF; [(μ).

i.e. $\angle$s DKG, DFE are equal;

$\therefore$ $\angle$s B, C are respectively equal to $\angle$s E, F.

Hence, two Triangles, which have their vertical angles equal, and the 2 sides of the one respectively equimultiples of those of the other, have their base-angles respectively equal.

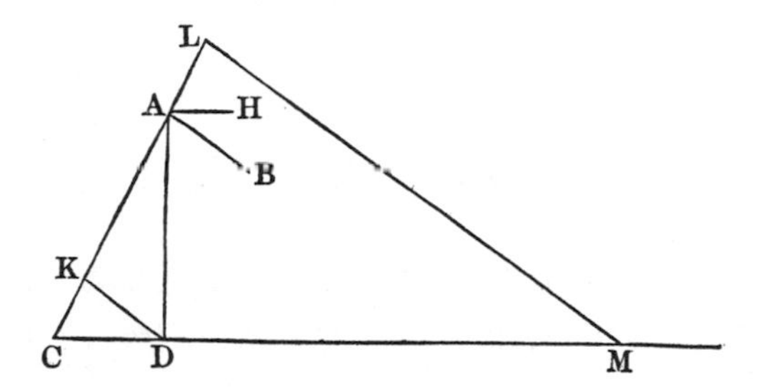

Now let AB, CD make with AC two interior $\angle$s BAC, ACD together $<$ 2 right $\angle$s. It shall be proved that they will meet if produced towards B, D.

In CD take any point D. Join AD. At A make $\angle DAH$ equal to $\angle CDA$.

Hence AH, CD are equally inclined to all transversals; [(μ).

$\therefore$ $\angle$s HAC, ACD together $=$ 2 right $\angle$s;

$\therefore$ they together $>$ $\angle$s BAC, ACD;

$\therefore$ $\angle HAC > \angle BAC$, i.e. $\angle HAD > \angle BAD$;

$\therefore$ $\angle CDA > \angle BAD$.

At D, in Line DA, make an $\angle$ equal to $\angle BAD$;

then the Line, so drawn, will fall within $\angle CDA$, and will meet CA between C and A. Call this point K.

In CA produced take CL a multiple of CK, and $>$ CA. And in CD produced take CM the same multiple of CD that CL is of CK. And join LM.

Because Triangles CKD, CLM have a common vertical $\angle$, and the 2 sides of the one respectively equimultiples of those of the other,

$\therefore$, by what has been already proved, $\angle$s CKD, CLM are equal.

Because AB, KD are equally inclined to AD,

$\therefore$ they are equally inclined to CL; [(μ).

$\therefore$ $\angle CAB = \angle CKD$ i.e. $= \angle CLM$;

$\therefore$ AB is separational from LM; [Euc. I. 28.

$\therefore$ if produced, it will meet CM.

Therefore a Pair of Lines, &c. Q.E.D.

APPENDIX IV.

List of Propositions of Euc. I, II, with references to their occurrence in the manuals of his Modern Rivals.

§ 1. References to Legendre, Cuthbertson, and Henrici.

Euclid	Legendre		Cuthbertson		Henrici	
I	I	Page	I	Page		Page
Ax 10	Ax 4	6	Ax	2	Ax 4	22
11	Th 1	,,	Th 8 Cor	17	Th	58
12	23	26			,,	72
1 Pr			Pr A	11		
2 ,,			(p)	74		
3 ,,			(q)	75		
4 Th	Th 6	10	Th 1	4	Th	129
5 ,,	12	14	2	5		
Cor	Cor	,,				
6 Th	Th 13	15	4	7		
Cor						
7 Th			(r)	76		
8 ,,	Th 11 Sch	14	5	8	Th	129
9 Pr	Pr 5	51	Pr B	12		
10 ,,	1	49	C	13		
11 ,,	2	50	7	16		
12 ,,	3	,,	11	20		

§ 1. References to Wilson, Pierce, and Willock.

Euclid		Wilson		Pierce		Willock	
I		I	Page		Page		Page
Ax 10		Ax 2	3	§ 16 Th	6		
11		Th	7			Th	10
12							
1	Pr					§ 1	33
2	,,					2	34
3	,,					3	36
4	Th	Th 16	26	§ 51,52	15	9	46
5	,,	8	21	55	16	1	40
	Cor	Cor	,,	56	17		
6	Th	10	22	58	,,	6 (I)	43
	Cor	Cor	,,	59	,,		
7	Th			60	18		
8	,,	18	28	61	,,	8	45
9	Pr	Pr 1	40	138	39	5	35
10	,,	3	42	132	38	6	36
11	,,	2	41	134	,,	7	,,
12	,,	4	42	133	,,	8	,,

EUCLID	LEGENDRE	Page	CUTHBERTSON	Page	HENRICI	Page
13 Th	Th 2	7	Th 9	18		
14 „	4	9	10	19		
15 „	5	„	6	14	Th	62
16 (*a*) „			13	22		
(*b*)			„	„		
17 „			„	23		
18 „	14	15	14	24	Th	108
19 „	„	„	15	25	„	„
20 „	8	11	16	26	„	109
21 „	9	12	17	27		
	II					
22 Pr			Pr D	30		
23 „	Pr 4	51	E	31		
	I					
24 Th	Th 10	12	Th 18	28	Th	133
25 „	Sch	13	19	29	„	„
26 (*a*) „	7	11	3	6	„	129
(*b*) „			25 Cor	43		
27 Th	Th 24 Sch	28	Th 20	36	Th	71
28 (*a*) „	„	„	„	37	„	„
(*b*) „	22	25	„	„	„	„
29 (*a*) „	24 Cor 2	28	21	38	„	„
(*b*) „	Sch	„	„	„	„	„
(*c*) „	24	„	„	39	„	„
30 „	25	29	22	40	„	72
	II					
31 Pr	Pr 6	52	Lemma	34		

Euclid	Wilson	Page	Pierce	Page	Willock	Page
13 Th	Th 1	8				
14 ,,	,,	,,	§ 24	8	Th 4	11
15 ,,	,,	9	23	7	2	10
16 (*a*) ,,	Th 7 Cor 4	20				
(*b*) ,,	,,	,,			12	13
17 ,,					,,	,,
18 ,,	9	21	62	18	6	43
19 ,,	11	22	,,	,,	6 (2)	,,
20 ,,	13	24			7	44
21 ,,	Th	35	40	12		
22 Pr	Pr 5	43	143	40	1	54
23 ,,	5 Cor 1	44	136	39	9	37
24 Th	Th 17	27	63	19	12	47
25 ,,	19	29			,,	,,
26 (*a*) ,,	15	26	54	16	10	46
(*b*) ,,					11	,,
27 Th						
28 (*a*) ,,						
(*b*) ,,						
29 (*a*) ,,						
(*b*) ,,						
(*c*) ,,						
30 ,,						
31 Pr						

Euclid	Legendre		Cuthbertson		Henrici	
	I	Page		Page		Page
32 (*a*) Th	Th 19 Cor 6	23	Th 24	42	Th	81
(*b*) ,,	19	20	,,	,,	,,	,,
Cor 1	20	23	30 Cor	49	,,	83
2			30	48	,,	,,
33 Th	30	32	23	41	,,	121
34 (*a*) ,,	28	30	26	44	,,	120
(*b*) ,,	,,	31	,,	,,	,,	,,
35 Th			Th 31	52		
36 ,,			Cor	53		
37 ,,			32 Cor	54		
38 ,,			,,	,,		
39 ,,			33	55		
40 ,,			Cor	,,		
41 ,,			32	54		
42 Pr			Pr L	59		
43 Th			Th 34	56		
44 Pr			Pr M. Cor	60		
45 ,,			N. Cor	61		
Cor						
46 Pr			Pr K	51		
46 Cor			Th 27	45		
	III					
			35	57		
47 Th	Th 11	71	36	58		
48 ,,						

Euclid	Wilson		Pierce		Willock	
		Page		Page		Page
32 (*a*) Th	Th 7	20	§ 71	21	Th 14 Cor 1	15
(*b*) ,,	,,	,,	65	20	14	14
Cor 1	6 Cor	16	72	21		
2	6	,,				
33 Th						
34 (*a*) ,,						
(*b*) ,,						
35 Th						
36 ,,						
37 ,,						
38 ,,						
39 ,,						
40 ,,						
41 ,,						
42 Pr						
43 Th						
44 Pr						
45 ,,						
Cor						
46 Pr						
46 Cor						
47 Th	29	63	256	74	14	84
48 ,,					18	88

Euclid		Legendre		Cuthbertson		Henrici	
II		**III**	Page	**II**	Page		Page
1	Th			Th 1	82		
2	,,			Cor 1	83		
3	,,			Cor 2	,,		
4	,,	Th 8	69	Th 2	84		
	Cor						
5	Th						
	Cor	10	70				
6	Th						
7	,,	9	,,	2	84		
8	,,			,,	85		
				I			
9	,,			Th (m)	72		
10	,,			Cor	,,		
				II			
11	Pr			Pr B	89		
12	Th	Th 13	74	Th 3	86		
13	,,	12	73	4	87		
14	Pr			Pr A	88		

Euclid		Wilson		Pierce		Willock	
II			Page		Page		Page
1	Th						
2	,,						
3	,,						
4	,,	Th 25	60			Cor 3	74
	Cor						
5	Th	27 Cor	62			10	82
	Cor	27	61			6	76
6	Th					11	83
7	Th	26	60			5	75
8	,,					9	78
9	,,	28	63			12	83
10	,,					13	84
11	Pr	Ex 3	67			14	104
12	Th	Th 31	66			16	87
13	,,	30	65			17	88
14	Pr	(4)	71	§ 289	84	3	96

§ 2. References to the other Modern Rivals.

Euclid	Chauvenet		Loomis		'Syllabus'-Manual	
I	I	Page	I	Page	I	Page
Ax 10	Ax	12	Ax 11	17	Def 5	7
11	1 Cor	14	1 Cor	18	Th 1	12
12	14 Cor 3	26	23 Cor 3	34		
1 Pr						
2 ,,						
3 ,,						
4 Th	20 Th	30	6 Th	22	Th 5	18
5 ,,	25 ,,	33	10 ,,	24	6	20
Cor	Cor	34	Cor 2	25	Cor	,,
6 Th	27 ,,	,,	11 ,,	,,	8	22
Cor	Cor	,,	Cor	,,	Cor	,,
7 Th						
8 ,,	22 ,,	31	15 ,,	27	15	28
	II		V			
9 Pr	28 Pr	78	5 Pr	95	Pr 1	61
10 ,,	25 ,,	76	1 ,,	93	4	63
11 ,,	26 ,,	77	2 ,,	94	2	62
12 ,,	27 ,,	,,	3 ,,	,,	3	,,
	I		I			
13 Th	2 Th	15	2 Th	19	Th 2	13
14 ,,	3 ,,	17	3 ,,	20	3	14
15 ,,	4 ,,	,,	5 ,,	,,	4	,,
16(*a*) ,,					9	22
(*b*) ,,					,,	,,
17 ,,					18	32
18 ,,	26 ,,	34	12 ,,	25	10	23
19 ,,	28 ,,	35	,, ,,	,,	11	24
20 ,,	17 ,,	29	8 ,,	23	12	25
21(*a*) ,,			9 ,,	24	13	26
(*b*) ,,	19 ,,	30				

§ 2. References to the other Modern Rivals.

Euclid	Morell		Reynolds		Wright	
I	I	Page	I	Page	I	Page
Ax 10	Ax	1			Ax	1
11	Th 1 Cor	5	Ax	2	Th	6
12	,, 25 Cor	26			,, 16 Sch	35
1 Pr						
2 ,,						
3 ,,						
4 Th	Th 5	9	Th 14	15	Th 8 (2)	16
5 ,,	,, 10	13	,, 11	12	,, 9 (3)	19
Cor	Cor 1	,,	Cor 2	,,	Cor	20
6 Th	,, 11	14	,, 12 Cor	13	,, 9 (1)	18
Cor	Cor	15			Cor	20
7 Th						
8 ,,	,, 9	12	,, 13	14	,, 8 (3)	16
					II	
9 Pr	Pr 9	66	Pr 2	30	Pr 9	87
10 ,,			,, 1	29	,, 8	86
11 ,,			,, 3	31	,, 11	88
12 ,,	,, 7	64	,, ,,	,,	,, ,,	,,
13 Th	Th 2	6	Th 1	3	Th 2	8
14 ,,	,, 3	7	,, ,,	,,	,, 3	,,
15 ,,	,, 4	8	,, 2	4	,, 5	11
16 (*a*) ,,			,, 10 Cor 2	12		
(*b*) ,,			,,	,,		
17 ,,						
18 ,,	,, 12	15	,, 12	13	,, 9 (4)	19
19 ,,	,, ,,	,,	,, ,,	,,	,, 9 (2)	18
20 ,,			,, 3	5	,,	13
21 (*a*) ,,	,, 8	12	,, 4	,,	,, 6	14
(*b*) ,,						

EUCLID	CHAUVENET		LOOMIS		'SYLLABUS'-MANUAL	
	II	Page	V	Page		Page
22 Pr	34 Pr	81			Pr 5	64
23 "	29 "	79	4 Pr	95	6	65
	I		I			
24 Th	24 Th	33	13 Th	26	Th 14	27
25 "	Cor	"	14 "	27	16	29
26 (*a*) "	21 "	31	7 "	23	7	21
(*b*) "					17	30
27 "	14 "	25	22 "	33	Th 21	43
28 (*a*) "	Cor 1	"	" "	"		
(*b*) "	" 2	26	21 "	32		
29 (*a*) "	13 "	24	23 "	33	22	44
(*b*) "	Cor 2	25			23 Cor	46
(*c*) "	" 3	"			"	"
30 "	12 "	24	24 "	34	24	"
	II		V			
31 Pr	30 Pr	79	6 Pr	96	Pr 7	66
	I		I			
32 (*a*) Th	18 Cor 1	29	27 Th	36	Th 25	47
(*b*) "	" Th	"	" "	"	"	"
Cor 1	29 "	37	28 "	37	26	48
" 2	Cor 2	38	29 "	"	Cor	49
33 Th	32 "	40	32 "	"	30	53
34 "	30 "	39	30 "	38	27, 28	50
					28	"
					II	
35 "					Th 1	82
36 "						
37 "						
38 "						
39 "					Th 2 Cor 3	84
40 "					"	"
41 "						

EUCLID	MORELL		REYNOLDS		WRIGHT	
		Page		Page	II	Page
22 Pr	Pr 6	64	Pr 7	32	Pr 6	83
23 ,,	,, 1	61	,, 4	32	,, 1	79
24 Th	Th 7	10			Th 7	15
25 ,,						
26 (*a*) ,,	,, 6	10	Th 15	16	,, 8 (I)	16
(*b*) ,,						
27 ,,	,, 25	25	,, 7 Cor 2	9	,, 16	33
28 (*a*) ,,	,, ,,	,,	,, 7	,,	,, ,,	,,
(*b*) ,,	,, ,,	,,	Cor 3	,,	,, ,,	,,
29 (*a*) ,,	,, 24	24	,, 8 Cor 2	10	,, ,,	,,
(*b*) ,,	,, 23	22	,, 8	,,	,, ,,	,,
(*c*) ,,	,, 24	24	Cor 3	,,	,, ,,	,,
30 ,,	,, 22	22	,, 9	11	,, 14 Sch	60
					II	
31 Pr	Pr 8	65			Pr 7	85
32 (*a*) Th	Th 28 Cor 1	28	Th 10	11	Th 19 Cor	41
(*b*) ,,	,, 28	27	,, ,,	,,	,, 19	,,
Cor 1	,, 29	29	,, 18	20	,, 20	42
2	,, 30	30	,, ,,	,,	Cor	43
33 Th	,, 33	33	,, 17	20	,, 22	48
34 ,,	,, 31	31	,, 16	19	,, 21	47
			and III 1	60		
			III		IV	
35 ,,			Th 2	61	Th 1	185
36 ,,			Cor	,,		
37 ,,			,, 4	62	Cor	186
38 ,,			Cor 1	,,		
39 ,,			Cor 2	63	Sch	187
40 ,,			,,	,,		
41 ,,			,, 3	62		

Euclid	Chauvenet		Loomis		'Syllabus'-Manual	
		Page		Page		Page
42 Pr					Pr 1	99
43 Th					Th 4	86
44 Pr					Pr 2	100
45 „						„
Cor						
46 Pr						
	IV		IV			
46 Cor						
47 Th	10 Th	133	11 Th	73	Th 9	91
48 „						
II					II	
1 Th					Th 5	87
2 „					Cor 2	88
3 „					Cor 1	„
4 „			8 Th	71	Th 6	„
Cor						
5 Th					8 Cor	91
Cor			10 „	73	8	90
6 Th					8 Cor	91
7 „			9 „	72	7	89
8 „						
9 „					13	96
10 „					„	„
11 Pr					Pr 6	103
12 Th	16 Th	113	13 Th	76	Th 10	94
13 „	15 „	112	12 „	75	11	95
14 Pr	12 Pr	136	22 Pr	103	Pr 4	101

Euclid	Morell		Reynolds		Wright	
		Page		Page		Page
42 Pr			Pr 1	76	Pr 2	198
43 Th			Th 1 Cor	60		
44 Pr						
45 ,,						
Cor						
46 ,,						
	V				IV	
46 Cor						
47 Th	Th 8	149	Th 12	69	Th 2	187
48 ,,						
II					IV	
1 Th						
2 ,,						
3 ,,						
4 ,,	Th 9	151	Th 9	67	Th 4	190
Cor						
Th					5	191
Cor	,, 11	152	,, 11	68	6 Cor	193
6 Th					,, 6	192
7 ,,	,, 10	152	,, 10	67	4 Sch	191
8 ,,						
9 ,,						
10 ,,						
11 Pr					Pr 3	198
	III		IV			
12 Th	Th 22	101	Th 17	97	Th 7	193
13 ,,	,, 21	100	,, 16	,,	8	194
14 Pr					Pr 2 Sch	198

A CATALOGUE OF SELECTED DOVER BOOKS IN ALL FIELDS OF INTEREST

A CATALOGUE OF SELECTED DOVER BOOKS IN ALL FIELDS OF INTEREST

AMERICA'S OLD MASTERS, James T. Flexner. Four men emerged unexpectedly from provincial 18th century America to leadership in European art: Benjamin West, J. S. Copley, C. R. Peale, Gilbert Stuart. Brilliant coverage of lives and contributions. Revised, 1967 edition. 69 plates. 365pp. of text.
21806-6 Paperbound $3.00

FIRST FLOWERS OF OUR WILDERNESS: AMERICAN PAINTING, THE COLONIAL PERIOD, James T. Flexner. Painters, and regional painting traditions from earliest Colonial times up to the emergence of Copley, West and Peale Sr., Foster, Gustavus Hesselius, Feke, John Smibert and many anonymous painters in the primitive manner. Engaging presentation, with 162 illustrations. xxii + 368pp.
22180-6 Paperbound $3.50

THE LIGHT OF DISTANT SKIES: AMERICAN PAINTING, 1760-1835, James T. Flexner. The great generation of early American painters goes to Europe to learn and to teach: West, Copley, Gilbert Stuart and others. Allston, Trumbull, Morse; also contemporary American painters—primitives, derivatives, academics—who remained in America. 102 illustrations. xiii + 306pp. 22179-2 Paperbound $3.00

A HISTORY OF THE RISE AND PROGRESS OF THE ARTS OF DESIGN IN THE UNITED STATES, William Dunlap. Much the richest mine of information on early American painters, sculptors, architects, engravers, miniaturists, etc. The only source of information for scores of artists, the major primary source for many others. Unabridged reprint of rare original 1834 edition, with new introduction by James T. Flexner, and 394 new illustrations. Edited by Rita Weiss. 6⅝ x 9⅝.
21695-0, 21696-9, 21697-7 Three volumes, Paperbound $13.50

EPOCHS OF CHINESE AND JAPANESE ART, Ernest F. Fenollosa. From primitive Chinese art to the 20th century, thorough history, explanation of every important art period and form, including Japanese woodcuts; main stress on China and Japan, but Tibet, Korea also included. Still unexcelled for its detailed, rich coverage of cultural background, aesthetic elements, diffusion studies, particularly of the historical period. 2nd, 1913 edition. 242 illustrations. lii + 439pp. of text.
20364-6, 20365-4 Two volumes, Paperbound $6.00

THE GENTLE ART OF MAKING ENEMIES, James A. M. Whistler. Greatest wit of his day deflates Oscar Wilde, Ruskin, Swinburne; strikes back at inane critics, exhibitions, art journalism; aesthetics of impressionist revolution in most striking form. Highly readable classic by great painter. Reproduction of edition designed by Whistler. Introduction by Alfred Werner. xxxvi + 334pp.
21875-9 Paperbound $2.50

THE PRINCIPLES OF PSYCHOLOGY, William James. The famous long course, complete and unabridged. Stream of thought, time perception, memory, experimental methods—these are only some of the concerns of a work that was years ahead of its time and still valid, interesting, useful. 94 figures. Total of xviii + 1391pp.
20381-6, 20382-4 Two volumes, Paperbound $8.00

THE STRANGE STORY OF THE QUANTUM, Banesh Hoffmann. Non-mathematical but thorough explanation of work of Planck, Einstein, Bohr, Pauli, de Broglie, Schrödinger, Heisenberg, Dirac, Feynman, etc. No technical background needed. "Of books attempting such an account, this is the best," Henry Margenau, Yale. 40-page "Postscript 1959." xii + 285pp. 20518-5 Paperbound $2.00

THE RISE OF THE NEW PHYSICS, A. d'Abro. Most thorough explanation in print of central core of mathematical physics, both classical and modern; from Newton to Dirac and Heisenberg. Both history and exposition; philosophy of science, causality, explanations of higher mathematics, analytical mechanics, electromagnetism, thermodynamics, phase rule, special and general relativity, matrices. No higher mathematics needed to follow exposition, though treatment is elementary to intermediate in level. Recommended to serious student who wishes verbal understanding. 97 illustrations. xvii + 982pp. 20003-5, 20004-3 Two volumes, Paperbound $6.00

GREAT IDEAS OF OPERATIONS RESEARCH, Jagjit Singh. Easily followed non-technical explanation of mathematical tools, aims, results: statistics, linear programming, game theory, queueing theory, Monte Carlo simulation, etc. Uses only elementary mathematics. Many case studies, several analyzed in detail. Clarity, breadth make this excellent for specialist in another field who wishes background. 41 figures. x + 228pp. 21886-4 Paperbound $2.50

GREAT IDEAS OF MODERN MATHEMATICS: THEIR NATURE AND USE, Jagjit Singh. Internationally famous expositor, winner of Unesco's Kalinga Award for science popularization explains verbally such topics as differential equations, matrices, groups, sets, transformations, mathematical logic and other important modern mathematics, as well as use in physics, astrophysics, and similar fields. Superb exposition for layman, scientist in other areas. viii + 312pp.
20587-8 Paperbound $2.50

GREAT IDEAS IN INFORMATION THEORY, LANGUAGE AND CYBERNETICS, Jagjit Singh. The analog and digital computers, how they work, how they are like and unlike the human brain, the men who developed them, their future applications, computer terminology. An essential book for today, even for readers with little math. Some mathematical demonstrations included for more advanced readers. 118 figures. Tables. ix + 338pp. 21694-2 Paperbound $2.50

CHANCE, LUCK AND STATISTICS, Horace C. Levinson. Non-mathematical presentation of fundamentals of probability theory and science of statistics and their applications. Games of chance, betting odds, misuse of statistics, normal and skew distributions, birth rates, stock speculation, insurance. Enlarged edition. Formerly "The Science of Chance." xiii + 357pp. 21007-3 Paperbound $2.50

EAST O' THE SUN AND WEST O' THE MOON, George W. Dasent. Considered the best of all translations of these Norwegian folk tales, this collection has been enjoyed by generations of children (and folklorists too). Includes True and Untrue, Why the Sea is Salt, East O' the Sun and West O' the Moon, Why the Bear is Stumpy-Tailed, Boots and the Troll, The Cock and the Hen, Rich Peter the Pedlar, and 52 more. The only edition with all 59 tales. 77 illustrations by Erik Werenskiold and Theodor Kittelsen. xv + 418pp. 22521-6 Paperbound $3.50

GOOPS AND HOW TO BE THEM, Gelett Burgess. Classic of tongue-in-cheek humor, masquerading as etiquette book. 87 verses, twice as many cartoons, show mischievous Goops as they demonstrate to children virtues of table manners, neatness, courtesy, etc. Favorite for generations. viii + 88pp. 6½ x 9¼.
22233-0 Paperbound $1.25

ALICE'S ADVENTURES UNDER GROUND, Lewis Carroll. The first version, quite different from the final *Alice in Wonderland,* printed out by Carroll himself with his own illustrations. Complete facsimile of the "million dollar" manuscript Carroll gave to Alice Liddell in 1864. Introduction by Martin Gardner. viii + 96pp. Title and dedication pages in color. 21482-6 Paperbound $1.25

THE BROWNIES, THEIR BOOK, Palmer Cox. Small as mice, cunning as foxes, exuberant and full of mischief, the Brownies go to the zoo, toy shop, seashore, circus, etc., in 24 verse adventures and 266 illustrations. Long a favorite, since their first appearance in St. Nicholas Magazine. xi + 144pp. 6⅝ x 9¼.
21265-3 Paperbound $1.75

SONGS OF CHILDHOOD, Walter De La Mare. Published (under the pseudonym Walter Ramal) when De La Mare was only 29, this charming collection has long been a favorite children's book. A facsimile of the first edition in paper, the 47 poems capture the simplicity of the nursery rhyme and the ballad, including such lyrics as I Met Eve, Tartary, The Silver Penny. vii + 106pp. 21972-0 Paperbound $1.25

THE COMPLETE NONSENSE OF EDWARD LEAR, Edward Lear. The finest 19th-century humorist-cartoonist in full: all nonsense limericks, zany alphabets, Owl and Pussycat, songs, nonsense botany, and more than 500 illustrations by Lear himself. Edited by Holbrook Jackson. xxix + 287pp. (USO) 20167-8 Paperbound $2.00

BILLY WHISKERS: THE AUTOBIOGRAPHY OF A GOAT, Frances Trego Montgomery. A favorite of children since the early 20th century, here are the escapades of that rambunctious, irresistible and mischievous goat—Billy Whiskers. Much in the spirit of *Peck's Bad Boy,* this is a book that children never tire of reading or hearing. All the original familiar illustrations by W. H. Fry are included: 6 color plates, 18 black and white drawings. 159pp. 22345-0 Paperbound $2.00

MOTHER GOOSE MELODIES. Faithful republication of the fabulously rare Munroe and Francis "copyright 1833" Boston edition—the most important Mother Goose collection, usually referred to as the "original." Familiar rhymes plus many rare ones, with wonderful old woodcut illustrations. Edited by E. F. Bleiler. 128pp. 4½ x 6⅜. 22577-1 Paperbound $1.25

TWO LITTLE SAVAGES; BEING THE ADVENTURES OF TWO BOYS WHO LIVED AS INDIANS AND WHAT THEY LEARNED, Ernest Thompson Seton. Great classic of nature and boyhood provides a vast range of woodlore in most palatable form, a genuinely entertaining story. Two farm boys build a teepee in woods and live in it for a month, working out Indian solutions to living problems, star lore, birds and animals, plants, etc. 293 illustrations. vii + 286pp.
20985-7 Paperbound $2.50

PETER PIPER'S PRACTICAL PRINCIPLES OF PLAIN & PERFECT PRONUNCIATION. Alliterative jingles and tongue-twisters of surprising charm, that made their first appearance in America about 1830. Republished in full with the spirited woodcut illustrations from this earliest American edition. 32pp. 4½ x 6⅜.
22560-7 Paperbound $1.00

SCIENCE EXPERIMENTS AND AMUSEMENTS FOR CHILDREN, Charles Vivian. 73 easy experiments, requiring only materials found at home or easily available, such as candles, coins, steel wool, etc.; illustrate basic phenomena like vacuum, simple chemical reaction, etc. All safe. Modern, well-planned. Formerly *Science Games for Children.* 102 photos, numerous drawings. 96pp. 6⅛ x 9¼.
21856-2 Paperbound $1.25

AN INTRODUCTION TO CHESS MOVES AND TACTICS SIMPLY EXPLAINED, Leonard Barden. Informal intermediate introduction, quite strong in explaining reasons for moves. Covers basic material, tactics, important openings, traps, positional play in middle game, end game. Attempts to isolate patterns and recurrent configurations. Formerly *Chess.* 58 figures. 102pp. (USO) 21210-6 Paperbound $1.25

LASKER'S MANUAL OF CHESS, Dr. Emanuel Lasker. Lasker was not only one of the five great World Champions, he was also one of the ablest expositors, theorists, and analysts. In many ways, his Manual, permeated with his philosophy of battle, filled with keen insights, is one of the greatest works ever written on chess. Filled with analyzed games by the great players. A single-volume library that will profit almost any chess player, beginner or master. 308 diagrams. xli x 349pp.
20640-8 Paperbound $2.75

THE MASTER BOOK OF MATHEMATICAL RECREATIONS, Fred Schuh. In opinion of many the finest work ever prepared on mathematical puzzles, stunts, recreations; exhaustively thorough explanations of mathematics involved, analysis of effects, citation of puzzles and games. Mathematics involved is elementary. Translated by F. Göbel. 194 figures. xxiv + 430pp. 22134-2 Paperbound $3.00

MATHEMATICS, MAGIC AND MYSTERY, Martin Gardner. Puzzle editor for Scientific American explains mathematics behind various mystifying tricks: card tricks, stage "mind reading," coin and match tricks, counting out games, geometric dissections, etc. Probability sets, theory of numbers clearly explained. Also provides more than 400 tricks, guaranteed to work, that you can do. 135 illustrations. xii + 176pp.
20338-2 Paperbound $1.50

POEMS OF ANNE BRADSTREET, edited with an introduction by Robert Hutchinson. A new selection of poems by America's first poet and perhaps the first significant woman poet in the English language. 48 poems display her development in works of considerable variety—love poems, domestic poems, religious meditations, formal elegies, "quaternions," etc. Notes, bibliography. viii + 222pp.
22160-1 Paperbound $2.00

THREE GOTHIC NOVELS: THE CASTLE OF OTRANTO BY HORACE WALPOLE; VATHEK BY WILLIAM BECKFORD; THE VAMPYRE BY JOHN POLIDORI, WITH FRAGMENT OF A NOVEL BY LORD BYRON, edited by E. F. Bleiler. The first Gothic novel, by Walpole; the finest Oriental tale in English, by Beckford; powerful Romantic supernatural story in versions by Polidori and Byron. All extremely important in history of literature; all still exciting, packed with supernatural thrills, ghosts, haunted castles, magic, etc. xl + 291pp.
21232-7 Paperbound $2.50

THE BEST TALES OF HOFFMANN, E. T. A. Hoffmann. 10 of Hoffmann's most important stories, in modern re-editings of standard translations: Nutcracker and the King of Mice, Signor Formica, Automata, The Sandman, Rath Krespel, The Golden Flowerpot, Master Martin the Cooper, The Mines of Falun, The King's Betrothed, A New Year's Eve Adventure. 7 illustrations by Hoffmann. Edited by E. F. Bleiler. xxxix + 419pp. 21793-0 Paperbound $3.00

GHOST AND HORROR STORIES OF AMBROSE BIERCE, Ambrose Bierce. 23 strikingly modern stories of the horrors latent in the human mind: The Eyes of the Panther, The Damned Thing, An Occurrence at Owl Creek Bridge, An Inhabitant of Carcosa, etc., plus the dream-essay, Visions of the Night. Edited by E. F. Bleiler. xxii + 199pp. 20767-6 Paperbound $1.50

BEST GHOST STORIES OF J. S. LEFANU, J. Sheridan LeFanu. Finest stories by Victorian master often considered greatest supernatural writer of all. Carmilla, Green Tea, The Haunted Baronet, The Familiar, and 12 others. Most never before available in the U. S. A. Edited by E. F. Bleiler. 8 illustrations from Victorian publications. xvii + 467pp. 20415-4 Paperbound $3.00

MATHEMATICAL FOUNDATIONS OF INFORMATION THEORY, A. I. Khinchin. Comprehensive introduction to work of Shannon, McMillan, Feinstein and Khinchin, placing these investigations on a rigorous mathematical basis. Covers entropy concept in probability theory, uniqueness theorem, Shannon's inequality, ergodic sources, the E property, martingale concept, noise, Feinstein's fundamental lemma, Shanon's first and second theorems. Translated by R. A. Silverman and M. D. Friedman. iii + 120pp. 60434-9 Paperbound $1.75

SEVEN SCIENCE FICTION NOVELS, H. G. Wells. The standard collection of the great novels. Complete, unabridged. *First Men in the Moon, Island of Dr. Moreau, War of the Worlds, Food of the Gods, Invisible Man, Time Machine, In the Days of the Comet.* Not only science fiction fans, but every educated person owes it to himself to read these novels. 1015pp. 20264-X Clothbound $5.00

9244